i

This Page Intentionally Left Blank

About Us

The Superintendent of the United States Military Academy (USMA) at West Point officially approved the creation of the Center for Nation Reconstruction and Capacity Development (C/NRCD) on 18 November 2010. Leadership from West Point and the Army realized that the United States Army, as an agent of the nation, would continue to grapple with the burden of building partner capacity and nation reconstruction for the foreseeable future. The Department of Defense (DoD), mainly in support of the civilian agencies charged with leading these complex endeavors, will play a vital role in nation reconstruction and capacity development in both pre and post conflict environments. West Point affords the C/NRCD an interdisciplinary and systems perspective making it uniquely postured to develop training, education, and research to support this mission.

The mission of the C/NRCD is to take an interdisciplinary and systems approach in facilitating and focusing research, professional practice, training, and information dissemination in the planning, execution, and assessment of efforts to construct infrastructure, networks, policies, and competencies in support of building partner capacity for communities and nations situated primarily but not solely in developing countries. The C/NRCD will have a strong focus on professional practice in support of developing current and future Army leaders through its creation of cultural immersion and research opportunities for both cadets and faculty.

The research program within the C/NRCD directly addresses specific USMA needs:
- Research enriches cadet education, reinforcing the West Point Leader Development Systems through meaningful high impact practices. Cadets learn best when they are challenged and when they are interested. The introduction of current issues facing the military into their curriculum achieves both.
- Research enhances professional development opportunities for our faculty. It is important to develop and grow as a professional officer in each assignment along with our permanent faculty.
- Research maintains strong ties between the USMA and Army/DoD agencies. The USMA is a tremendous source of highly qualified analysts for the Army and the DoD.
- Research provides for the integration of new technologies. As the pace of technological advances increases, the Academy's education program must not only keep pace but must also lead to ensure our graduates and junior officers are prepared for their continued service to the Army.
- Research enhances the capabilities of the Army and DoD. The client-based component of the C/NRCD research program focuses on challenging problems that these client organizations are struggling to solve with their own resources. In some cases, USMA personnel have key skills and talent that enable solutions to these problems.

For more information please contact:

Center for Nation Reconstruction and Capacity Development
Attn: Dr. John Farr, Director
Department of Systems Engineering
Mahan Hall, Bldg. 752
West Point, NY 10996
John.Farr@usma.edu
845-938-5206

This Page Intentionally Left Blank

Table of Contents

List of Figures

List of Tables

Abbreviations and Acronym List

Abbreviation /Acronym	Explaination
AVG	Average
CAES	Compressed-air energy storage
CES	Community energy storage
DoE	Department of Energy
DoD	Department of Defense
EV	Electric vehicles
ft^2	Square feet
kW	Kilowatt
KWh	Kilowatt hour
LEED	Leadership in Energy and Environmental Design
MW	Megawatt
NG	Natural Gas
NREL	National Renewable Energy Laboratory
NY	New York
OSI	Open Systems Interconnection
PHEV	Plug-in hybrid electric vehicles
RTO	Regional Transmission Operator
U.S.	United States
USMA	United States Miitary Academy

This Page Intentionally Left Blank

Chapter 1
Introduction

1.1 Background

The United States (U.S.) is in a state of extreme dependence on fossil fuels often from reliable nations to support our economic and social well being. The government and other stakeholders have not only begun to understand the damage that fossil fuels have on the environment, but also to understand that this dependence places the U.S. at an extreme risk to a host of attacks and embargos. Thus, the Army Energy Security Implementation Strategy and Campaign Plan[1] was conceived to "ensure that the Army provides safe, secure, reliable, environmentally compliant and cost-effective energy and water services to soldiers, families, civilians and contractors on Army installations." Also numerous mandates, orders, and laws have been decreed that dramatically effect energy and environmental issues within the Army. For example, on the Army Energy Program website[2] 14 Army Guidelines, 4 Department of Defense (DoD) Guidance, 5 Presidential Orders, and 5 Federal Laws and Statutes are referenced as affecting Army energy and environmental issues. As shown in Figure 1.1, the modern installation manager and commander must navigate a complex environment to ensure that an installation can conduct its mission both in war and peacetime.

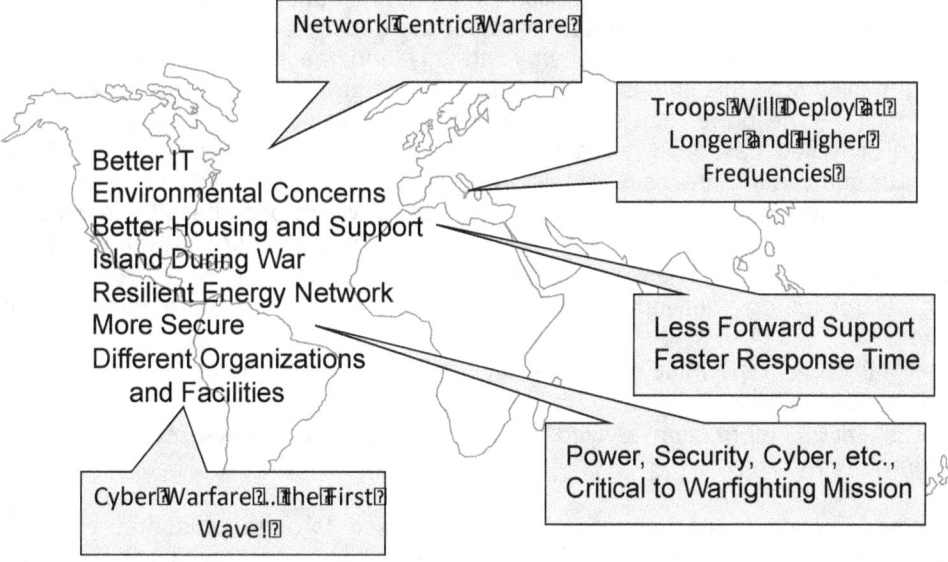

Figure 1.1 Issues for the 21st century military installation

If the U.S. were attacked from either a sophisticated enemy or even a disgruntled employee, the energy grid would be an easy target with catastrophic results. A recent New York Times article stated that the People's Liberation Army of China is growing corps of cyberwarriors and "has drained terabytes of data from companies like Coca-Cola, increasingly its focus is on companies involved in the critical infrastructure of the U.S. — its electrical power grid, gas lines and waterworks. According to the security researchers, one target was a company with remote access to more than 60 percent of oil and gas pipelines in North America."[3] This would cut power off to military bases and would greatly degrade the

---

[1] See http://www.asaie.army.mil/Public/Partnerships/doc/AESIS_13JAN09_Approved%204-03-09.pdf accessed 14 November 2011
[2] See http://army-energy.hqda.pentagon.mil/policies/key_directives.asp accessed 14 November 2011
[3] See http://www.nytimes.com/2013/02/19/technology/chinas-army-is-seen-as-tied-to-hacking-against-us.html?partner=rss&emc=rss&src=igw&_r=0 accessed 19 February 2013

force projection capability of the military installation along with crippling the U.S. economy. Creative economics and engineering solutions are needed to finance the upgrades needed not only to ensure the security of our installations, but also to comply with the myriad of regulations, orders, and laws. Lastly, research is needed into taxonomy and complex systems dependences between installation and regional energy, dependencies of operational and installation energy, and a common taxonomy for the Army to use when discussing energy.

The electric power industry or system is often split up into four processes: generation (i.e., power stations), power transmission, distribution and retailing. The motivation of cost, environmental impact, the war on terrorism and a growing population has increased the scrutiny of electrical power system in the U.S. A reengineering of the electrical system along the lines of the Internet could yield potentially significant benefits. The architecture of communication over distributed networks is currently thought of in the context of a set of layers such as the OSI 7 layer model[4]. These models help to efficiently decompose route and transmit information according to the needs or constraints of the network participants. Such approaches yield robust and scale-able systems. A similar perspective can be leveraged in the energy field as well. With the advent of benefits from smart grid technology a formal layered approach for energy generation, transmission, and consumption would be necessary and beneficial.

The important differences between information and energy transmission would necessarily yield a significantly different architecture. However, many of the concepts and approaches would be similar. Figure 1.2 shows the potential shift in energy strategy and the means by which this shift can be accomplished. Implied from this shift is that consumer demand is central to the network and is typically expressed in pricing. The power grid will act similarly to a data network in that particular power loads can be applied or routed based on demand, either in preference or power source. This demand is random in nature and can include issues dealing with reliability, sustainably, as well as cost and energy efficiency. The essence of an energy network is that routing will be price based and dynamic. This is a significant departure from a switch-based network. There are many hurdles with this type of problem that should be explored further. These include but are not limited to:

- Is a layered energy network architecture an evolution or departure from the current power system?
- What is the role of the value of information in such a routing scheme?
- Can consumers become producers to the system? and
- How is pricing information relayed through the system to drive investments and decision-making?

These and other important questions should be addressed through collaborative research efforts. Implementation of an energy network architecture would allow for improved energy security with the benefits of cost and efficiency improvements.

---

[4] The Open Systems Interconnection or OSI model is used for characterizing and standardizing the functions of a communications system in terms of abstraction layers.

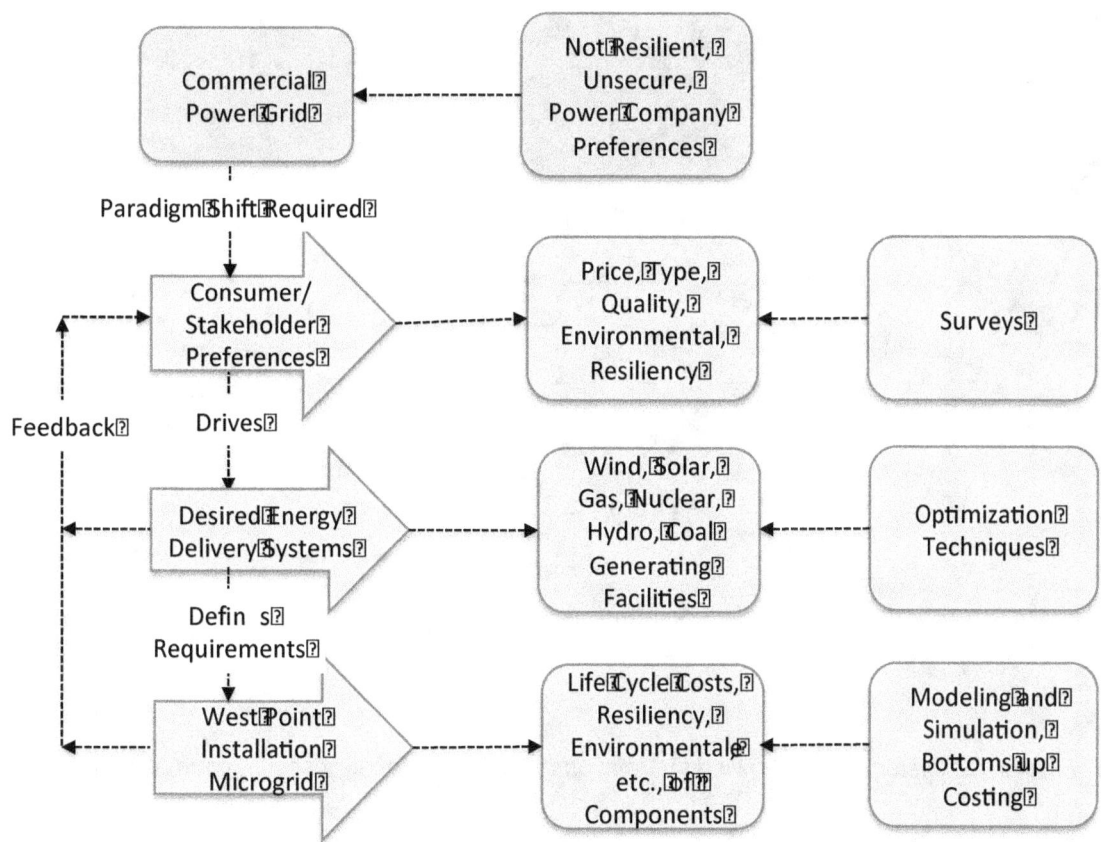

Figure 1.2 *Vision for energy price based routing strategy*

1.2 Problem Statement

Figure 1.3 shows the interaction of NetZero and energy security for the electric power system. However, as shown in Figure 1.4, NetZero and security are only a portion of the challenges that must be addressed as part of the Army's operational energy strategy. Thus, this project will develop new energy pricing and buying strategies with key tasks of reducing budgets, creating Netzero installations, and improving energy security. Potential solutions to this problem would include the development and implementation of smart grids, micro grids, new technologies, behavior modification, or a combination of these technologies.

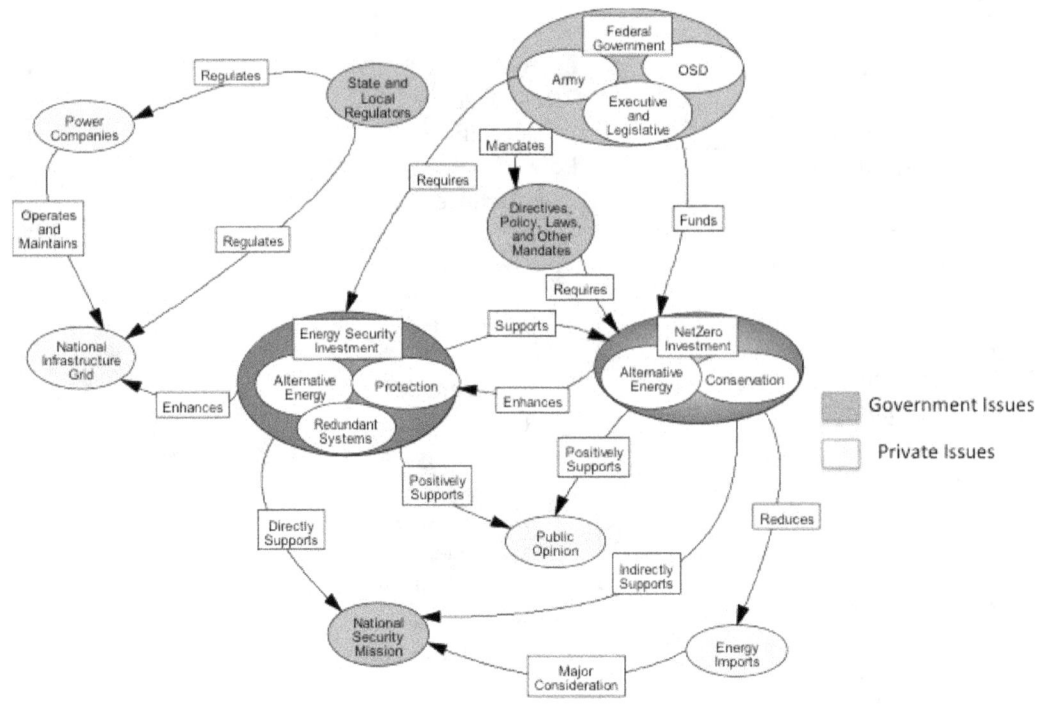

Figure 1.3 *Systemigram showing the relationship between energy security and NetZero*

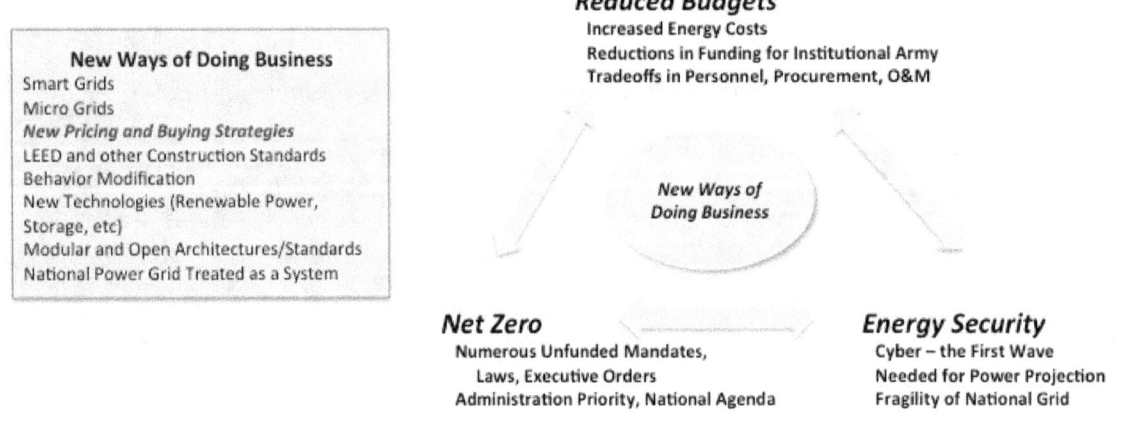

Figure 1.4 *The Army's operational energy strategy*

The systemigram in Figure 1.5 encapsulates the scope of this report by illustrating the relationships between energy characteristics, energy solutions, and energy technology. Smart grids and micro grids represent the emerging technologies that support alternative solutions to the growing concerns of security and resource depletion. These solutions, paired with government policies and consumer preferences impact the energy characteristics that create various energy profiles. This report will focus on creating a methodology for an economic solution that influences changes to the system through investing in new energy sources and consumer preference maximization.

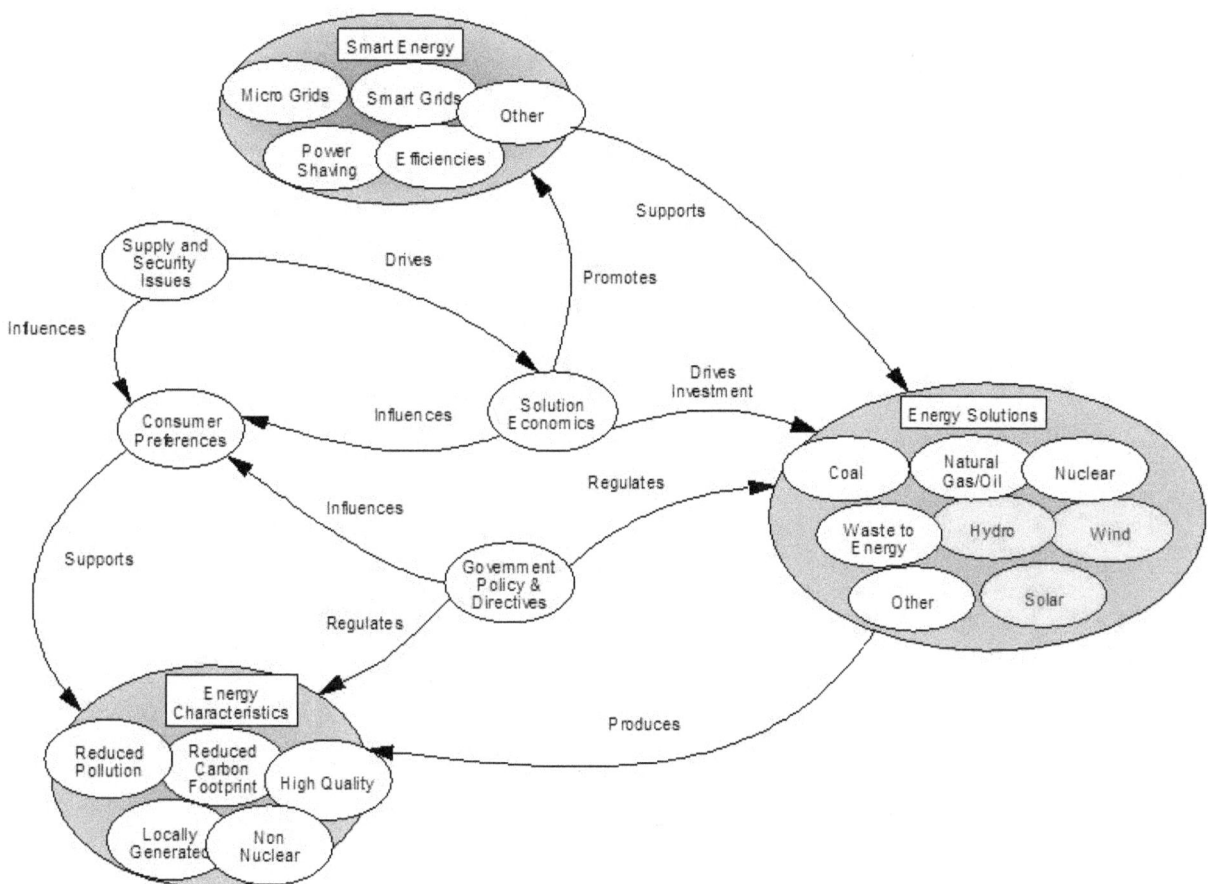

Figure 1.5 *Systemigram showing the relationship between solutions, policy, and preference*

This Page Intentionally Left Blank

Chapter 2
Literature Review

2.1 Micro/Smart Grids

An important concept to understand throughout this report is called a micro grid, which are "small electrical distribution networks that can be operated in islanded mode or interconnected with the mains" (Braun and Strauss, 2008); the key point being that a micro grid can be self sufficient. Micro grids can pull power from the main grid during peak times if needed, but can mostly self sustain itself. The energy market in developed nations mostly relies on macro grids. Treating the energy market as a system, there currently are many concerns associated with the current structure. When dealing with such large areas, the macro grid makes it difficult the forecast consumer demand and produce the required energy. Another disadvantage of the macro grid is that it monopolizes power. This leads to higher prices for the consumers (Marnay and Venkataramanan, 2006).

Arguments can be made that a move to micro grids from traditional large power distribution networks, macro grids, would be a return to the past. The first power grids created were often very isolated from one another. Each was self sustaining with no reliance on other grids. One of the first examples of a micro grid was Thomas Edison's Manhattan Pearl Street Station in 1882. However, the era of isolated systems was short lived in large part due to advances in alternating current, ac, technology. The first system to utilize ac technology was implemented in 1896. This system carried electricity from a hydro station in Niagara Falls to Buffalo, NY (Marnay and Venkataramanan, 2006). Current examples of micro grids can be seen in the island of Lemnos and in the English Borough of Woking. Lemnos uses wind turbines, solar panel, as well as diesel generators to create 14.84 MW of power. Woking uses different forms of energy production including combined heat and power plants and fuel cells to generate 2 MW of power (Hernandez-Aramburo et al., 2005).

Utilization of micro grids offer many advantages such as cost reduction for the consumer, renewable resource utilization, improvement in reliability, and a reduction in negative effects of the environment due to existing power generation (Marnay and Venkataramanan, 2006). As stated earlier, the micro grid in Lemnos uses wind turbines and solar panels to create energy. This is effective in a micro grid due to the size of the system. Many of these renewable resources are not currently able to generate late wattages based on the current technologies. This makes it difficult for a wind turbine plant to supply enough energy for a large geographic region. However, utilization of wind power in a small town is much more feasible considering this micro grid would be able to meet consumer energy demands. Micro grids can also increase reliability in the system. They have the potential to not only turn on when local supply needs more energy, but micro grids can also be used to help reduce an excessive load in the macro grid. The utilization of micro grids during these peak times would decrease the potential for blackouts (Eto et al., 2000). Not only can micro grid provide all of the electricity and heating needs for local customers, but it can also reduce overall emissions during the energy production process. Micro grids make up a part of a larger, more effective power supply and distribution network known as Smart Grids.

The U.S. Department of Energy (DoE) defines a smart grid as the following:

"A smart grid uses digital technology to improve the reliability, security, and efficiency of the electricity system, from large generation through the delivery systems to electricity consumers. Smart grid deployment covers a broad array of electricity system capabilities and services enabled through pervasive communication and information technology, with the objective of improving reliability, operating efficiency, resiliency to threats, and our impact on the environment." (Department of Energy, 2012)

The Smart Grid tailors to the demands of the 21st century electricity consumers by optimizing energy efficiency and combating new security challenges. In order to accomplish this, a Smart Grid has to rely on new digital technology, specifically focused on communication between the consumer and the producer.

There are six major characteristics of a smart grid (Department of Energy, 2012):
1. Enables informed participation by customers
2. Accommodates all generation and storage options
3. Enables new products, services and markets
4. Provides power quality for the range of needs
5. Optimizes asset utilization and operating efficiency
6. Operates resiliently to disturbances, attacks and natural disasters

The concept of the smart grid arose from the need to diversify energy resources to promote energy security and meet the demands of consumers. As more developments emerged, the scope of the smart grid expanded to drive a new electrical system. These drivers include the environment, system reliability and operational excellence (Rahimi, 2010). These drivers push the smart grid into being adaptable, to better meet the new demands of the consumers and environmental challenges. The success of a smart grid system relies on meeting changing and evolving demands and solving a number of implementation problems and factors.

The smart grid works as a multi-way communication system integrating the consumer into the distribution grid, ultimately communicating with the supplier. In order for the success of this system, the consumer needs an interface in which they can communicate with the network. The suppliers also need to be able to respond to these changing demands quickly as well as forecast future demands in order for them to meet power requirements. The power supply derives its power from not just one main power source but rather many distributed sources in order to better meet individual demands and increase reliability in the system. This would also decrease transmission losses, one of the largest losses in the electric grid. In the past, power systems were based off huge power generation systems such as a coal power plant, and distributed through a traditional hub and spoke method. This centric system caused power to be transported over long distances to reach consumers within the region. The transportation of energy created large losses of electricity through transmission and increased the overall price for electricity. If this distribution system failed, for example a tree falling on power lines, there was no way to reroute this power and customers had to wait until the power crews restored the power lines and eventually electricity. This has significant impacts on power-dependent industries and business such as grocery stores, hospitals (who usually own their own generators), and factories.

In the smart grid, power generation will also retain the traditional system while also adopting smaller power generators distributed throughout the system. These power generators integrate within a new distribution system, bringing more reliable power and reducing transmission costs. Power generation sources would be more evenly distributed throughout the system as the prevalence of windmills and solar panels increase. Energy could be supplied through different points within the system, acting as a failsafe in case a power generator such as a coal power plant fails. Consumers would also be able to better manage the type and amount of power they would be receiving, meeting their interests and creating a market for certain types of power. For example, one consumer might prefer carbon neutral power and be willing to pay more for it. Power suppliers would then meet this market demand and produce more carbon neutral power to take advantage of the economic surplus. The fundamental driver behind the functionality and management of the smart grid is an advanced distribution management system.

The advanced distribution management system for the smart grid starts with data acquisition and supervisory control (Fan, 2009). The current dispatch and system operation systems operate with insufficient or minimal data and on manual or analog systems. Operators have to work with multiple interfaces and systems, restricting the optimization and efficiency of the system. Furthermore, the experience of the operators is critical for the safety of the system. By creating a new supervisory control and data acquisition system, costs can be reduced while efficiency, reliability and asset utilization can be maximized (Fan, 2009). With an advanced distribution management system and a consumer interface, the power grid can meet the changing demands of consumers while maximizing cost and operational efficiency.

Demand response underlies the foundation of smart grids and remains one of the most important elements in the smart grid system. Demand response can be defined by the actual use of electricity by consumers in response to changes in pricing controls or other direct and indirect methods to change consumer consumption patterns (Fuller, 2012). Demand response strives for effective and efficient utilization of power sources and assets (Rahimi, 2010).

As consumers get integrated into a distributed management system, the power generation and the distribution network have to be able to successfully communicate with the consumer and meet their power demands. This dynamic change illustrates the challenges that load management has faced since the early 1980s (Ipakchi, 2009). Load management takes power strategies such as peak shifting and direct load control and implements them to manage power supplies so producers are able to meet the high demands of consumers. With demand response, system operators are able to use control methods that consumers can respond to, changing their level of demand. This changes the dynamic of the load whereby system operators will be able to monitor and control the consumer demanded load through direct or indirect methods such as pricing controls to meet electricity needs and maximize the utilization of assets.

2.2 Electricity Pricing

Based on the ever evolving growth of new technologies such as the smart grid, micro grid, and renewable energy production new pricing strategies must be developed to capture the dynamics of supply and demand in this market. This electricity pricing strategy development is one of the most important and complex aspects of this dissertation. The dynamic pricing strategies used to offer customers shifting prices depend upon several internal and external factors. Dynamic pricing makes value and cost of energy use transparent to consumers which enable them to determine when cost exceeds value thus enabling the alteration of their energy consumption and production needs (Kiesling, 2008). As smart grid and other technologies continually emerge it is also necessary to develop an efficient means of pricing. There has been an array of attempted pricing and forecasting models constructed in an attempt to increase end use efficiency, energy conservation, capacity utilization, savings, and to insure energy security. Crucial to understanding the theory behind a price based network architecture is an understanding of marginal costs, consumer supply and demand, and the effects of deregulation and competition within a market. Based on research of several pieces about the intricacies and logistics of electricity pricing the following factors, concerns, and existing models will be examined during the formulation of a new architecture and pricing model for smart grid and micro grid technologies.

As with any complex system there are many different factors that impact electricity pricing, even with advances in technological capacity there are factors that affect pricing schemes no matter which model is implemented. Prices fluctuate by city, state, region and input source (petroleum, oil, coal, etc.) (Bureau of Labor Statistics, 2012). Other factors that affect the price of electricity and energy include:
- peak and demand as rate payers share the cost for an electric power company to make capital investment in order to meet anticipated peak demands;
- a limited use of alternative energy technology despite its advances in recent years because of implementation costs, and state and legal constraints;
- the state and local taxes imposed on utilities;
- the power generation or input source;
- environmental considerations that make it more expensive for companies to use technology with a smaller carbon footprint; and
- the type of transmission used, more specifically losses over distance traveled and the installation of transmission technology (Banks, 1984).

These factors also serve as concerns that spark a need to minimize the costs of these factors as well as insuring the efficient transmission of electricity and energy. The current market is undulating between monopolies of major regional power providers to deregulation and competition. In the emerging deregulated market customers have the power to choose their provider and save on energy (NY Electric Rates, 2012). This competition, paired with advances in technology which provides the means for large scale implementation (Fahey, 2010), gives clients at all levels "the opportunity to compare suppliers' rates, decide who best fits their energy consumption needs, choose providers, and save money" (NY Electric Rates, 2012). With the emergence of new technology and emerging markets there is a pressing need for accurate forecasting models and pricing schemes. Currently, there are several models used to keep up with the "electricity infrastructure [that] has been transformed into progressively competitive electricity industries" (Larsen and Bunn, 1999).

The emerging forecasting models are in response to the need to forecast prices in the competitive electricity market for both producers and consumers.(Nogales and Conejo, 2006) This need is due to the "requirement for a safe and efficient electricity grid" where "supply and demand [are] always in perfect balance"(Ramchum et al., 2012). To acquire this harmony, models employ demand side management, static time of use pricing, transfer pricing models, critical peak pricing, locational marginal prices, and marginal cost forecasting techniques to insure supply meets demand. Each technique possesses a handful of advantages and disadvantages. Demand side management varies the supply side in real time to match demand but initiatives to reduce demand tend to "reduce the natural diversity of consumers' peak demands and shift all of those peaks to specific periods" which raises a concern of disrupting the grid (Ramchum et al., 2012). In the static time of use pricing scheme model "utilities charge rates that depend on the time of day and the season of year at which electricity is used" (Spector et al., 1995). With this model, individuals and firms can adjust their output rates based on peak, mid peak, and off peak times of the day and year. One of the concerns with this model is the significant additional peaks in demand as soon as the off peak period is reached and individuals and firms restart their electricity and energy consumption (Ramchum et al., 2012). Transfer pricing models provide predicted pricing information at some time horizon before the rates will apply that affects the use and purchase of certain types of electricity at certain times. This model uses electricity demand as an explicative variable which improves predictions (Nogales and Conejo, 2006). However, it also has nonconsistent mean and variance, high frequency, high volatility, the presence of outliers, daily and weekly seasonality, and a calendar affect on weekends and holidays; (Nogales and Conejo, 2006) all of which introduce error and complexity into the model. Critical peak pricing focuses on the peak demand times throughout the day to influence energy pricing, during the critical period prices are higher which is very similar to the time of use pricing model; a drawback of this approach is again the additional peaks created as devices turn back on as soon as the critical period is over (Ramchum et al., 2012). "In the deregulated power system, locational marginal prices are used in transmission engineering predominantly as near real time pricing signals" (Sathyanarayana, 2012). This model depicts the direction of market prices and the cost of satisfying the next increment in energy demand for a certain node (Sathyanarayana, 2012). Implemented to optimize system operation, this pricing scheme is focused on the advent of smart grid technology by developing control signals used for pricing power and energy at ultimate points of delivery (Sathyanarayana, 2012). A final model researched was the marginal cost pricing model which yields less capital requirements, better capacity utilization and higher end use economic efficiency (Guldmann, 1986). This model includes sub models for demand, supply, and a procedure that determines equilibrium prices iteratively (Guldmann, 1986).

Yet, these models fall short with respect to smart grid and micro grid technology architectures. The pricing schemes do not keep pace with the advent of rapidly changing technologies. Pair the inadequacies of past pricing models to accurately forecast electricity supply and demand in an attempt to mitigate complications from peak demands and overconsumption with ever evolving initiatives for cost efficient, 'green' energy and there is clearly a problem. The contrast between a shrinking budget for the Institutional Army or the economic recession occurring in America and increasing energy costs indicates a need to be more cost efficient and to lessen the losses or inefficiencies present in the current system.

In order to keep pace with new technologies and meet important environmental goals it is necessary to adapt to new ways of doing business within the energy grid sector.

There are several growing concerns that stem this need. The development of new technologies that allow for a potential large scale implementation of smart grids and micro grids yield the requirements of new energy system architectures. The national power grid, instead of a monopoly dominated by a handful of large power distributors will need to be treated as a unique system with complex endogenous and exogenous variables that impact the overall market. As new ways of doing business grow into a requirement instead of a choice there is a need to develop new pricing and buying strategies and models that will accurately and efficiently model emerging architectures. There are several requirements of this new pricing strategy that will function with price as the driving factor behind energy production and consumption – a price based energy network architecture.

The pricing strategy must improve the cost efficiency of the production, transmission, and consumption of electricity in order to combat reductions in budget. It must also take into account the need to meet environmental initiatives and ensure the secure transmission and reception of electricity across the grid. To meet the pricing strategy necessities for smart grid technology many researchers have indicated the need to implement dynamic pricing strategies to illicit demand response to electricity prices (Faruqui and Sergici, 2010). Demand response is the alterations of consumer activity with respect to energy consumption based on outstanding factors – in this case electricity pricing (Faruqui and Sergici, 2010) Dynamic pricing attempts to provide current supply and demand situation depending upon the date of delivery for a price which depends on the time and location of use (Faruqui and Sergici, 2010). A key question linking these two phenomena is whether or not dynamic pricing can be the driving factor behind the demand response of energy consumers? Would dynamic pricing return the greatest value on the investment into smart technology and a new electrical grid structure?

According to economic researchers, "any structural model for the industry [electricity] should include a mechanism for charging consumers for the cost of the production and delivery of electricity at the time of its consumption" (Faruqui and Sergici, 2012). This is the crux of dynamic pricing and it is an extremely important aspect that should be implemented effectively to keep pace with the ever changing technology used in America's electric grid. "If you're going to have a smart grid, that allows you to measure and have two-way communication between the end use premises, the utility company, the RTO (Regional Transmission Operator), and other entities, rates will have to change to be more time of use rates or critical peak period rates" (Faruqui and Sergici, 2010). The alterations in technology make alterations in pricing strategy a pressing concern in the field.

These new pricing strategies can also serve as the key variable which drives energy demand and efficient use as "consumers will respond to higher prices by lowering usage" (Faruqui and Sergici, 2010) The magnitude of this response, as seen through tests conducted by the Brattle Group in California, depends upon these factors: "magnitude of price increase, presence of air conditioning, and the availability of enabling technologies such as two way programmable communicating and always on gateway systems that allow multiple end uses to be controlled remotely" (Faruqui and Sergici, 2010). It has been shown, through pilot experiments that customers respond to dynamic pricing by altering their consumption around peak and off peak time periods, customer response persists over time indicating a potential long term change in behavior, and customers respond to information feedback on utility bills, energy uses, and *prices* (Faruqui and Sergici, 2011). A representation of the data collected on the reduction in demand based on the experiments conducted can be seen in Figure 2.1.

Photo Removed Due to Copyright Restrictions

Figure 2.1 Reduction in demand in response to various pricing pilots (Faruqui and Sergici, 2010)

The change is quantifiable with the reduction of energy use during critical peak time periods in response to time of use or critical peak dynamic pricing schemes. In an experiment conducted by the Brattle group investigating the household response to such pricing schemes time of use rates induced a drop in peak demand that ranged "between three to six percent" and critical peak pricing tariffs induced a drop in peak demand that ranged "between 13 to 20 percent. When accompanied with enabling technologies, the latter set of tariffs leads to a drop in peak demand in the 27 to 44 percent range" (Faruqui and Sergici, 2010). This study is an indication of the impact pricing can have on developing new ways of doing business to ensure the sustainability and cost efficient use of a new price based electricity grid architecture. "The demand responses vary from modest to substantial due to a variety of factors, some of which can be controlled such as electricity prices and whether or not enabling technologies are present, and some of which cannot be controlled, such as the design of the experiment and its location" (Faruqui and Sergici, 2010). Despite variation in response, there is clearly a link between changing prices and consumption, with that in mind this paper will investigate new pricing and buying strategies for a price based energy network architecture.

Based on the inefficiencies of current pricing models due to the advent of new technologies such as the smart grid and micro grid there is a pressing need to develop new ways of doing business that are cost efficient to meet reduced budgets, environmentally friendly, and ensure secure transmission of energy across the grid. These new ways of doing business require new pricing and buying strategies that will be researched for feasibility and effectiveness in this dissertation.

2.3 The Influence of Government Policy and Strategy

The link between policy and energy is a critical aspect of developing new methods for the electrical grid. Government policy has the power to regulate the types of energy sources utilized in the market through various directives as well as the different characteristics produced by those sources. Policies can also influence consumer preferences based on different directives and initiatives.

Government policy and other political factors significantly affect energy regulation and energy industries (Chang and Berdiez, 2011). Policies can be negatively enforced or positively enforced to help prevent or promote different results. For example, taxes and regulations could be used to reduce pollution and encourage the development of new technologies by shifting behaviors (Gerlagh, 2008). The different policies and directives can be inspired by various influences. The political landscape certainly influences the types of energy used within a country as well as any advancements to the grid through economic incentives – these advancements due to economic incentivizing is called induced innovation; induced innovation has a substantial role in reducing costs for energy and increasing research into different sources of production (Gerlagh, 2008). In more general terms, government policy has consequences on pricing and production – changes in the capacity and generation of energy are driven by price (Paul et al., 2013). These changes are ultimately shaped by government policy and initiatives that greatly influence not only electricity producers but also consumers (Paul et al., 2013).

The impact of government policy and directives such as NetZero is significant and drives social behaviors, business practices, and energy initiatives for the national energy grid as well as the DoD; yet, it is possible for the relationship to be symbiotic. The energy sector can help to influence government policy at times creating an interactive relationship rather than a reactive relationship. While government leaders can develop strategies and initiatives that focus on securing energy, protecting the environment, and reducing emissions to alter climate change it is incumbent upon research to develop solutions to meet the vision and put it into practice (Obama, 2013). Considering the looming energy crisis, depleting natural resources, and pending threats of natural disasters or attacks it is clear that there ought to be, and is, a focus on altering the energy grid to mitigate these concerns. There are several different alternatives to developing answers to these pressing questions with the objective of building a stronger, smarter, and cleaner electric grid that increases the efficiency of our energy grid (Obama, 2013). This report focuses on creating an economic solution that is commensurate with government initiatives and helps to drive investment into new energy sources by uncovering economic inefficiencies.

The basis of pricing as a means to influence consumer preferences and energy source investment can also factor into government policy. By influencing behavior in the social and business sectors it will help reinforce government policies focused on advancing the energy grid. The economic inefficiencies in the market uncovered by this report's methodology may put pressure on politicians to alter their ideology that significantly "influence[s] policy proposals" with regards to energy (Chang and Berdiez, 2011). Further, pricing policy and strategy can provide incentives (Paul et al., 2013) for producers to invest in research and development into new technology which in turn alters buying practices for consumers with more choices to meet their preferences.

The relationship between government policies, initiatives, and directives and the energy grid is a critical factor in developing a methodology to optimize efficiency and mitigate many of the concerns facing the U.S. and the DoD at this time. With a looming energy crisis and other threats to the grid energy is and will continue to be a pressing policy issue. Understanding the relationships between policy and energy sources and outputs is imperative to building economic solutions to the issue. These solutions should focus on meeting policy goals and help to drive further policy development by way of providing economic incentive for change through uncovering vast economic inefficiencies that need to be remedied to help create more efficient and safe grid for producers and consumers.

2.4 Renewable Energy Storage

The reliability and operational flexibility of the electric distribution system depends largely on the effective application of renewable energy and storage devices. Renewable energy storage lets electric energy producers send excess electricity over the electricity transmission grid to temporary electricity storage sites that become energy producers when electricity demand is greater, optimizing the production by storing off-peak power for use during peak times. Many renewable energy technologies such as solar and wind energy cannot be used for base-load power generation as their output is much more volatile and depends on the sun, water currents or winds. Batteries and other energy storage technologies therefore become key enablers for any shift to these technologies. The power storage sector generally includes traditional batteries, but also covers hydrogen fuel cells and mechanical technologies like flywheels that are straight potential replacements for batteries. More and more research is also conducted in the field of nanotechnology as ultra-capacitors (high energy, high power density electrochemical devices that are easy to charge and discharge) and nano-materials could significantly increase the capacity and lifetime of batteries.

The replacement of fossil fuels as the primary sources of energy for electricity generation and transportation needs to take place over the next few decades. Growing penetration of renewable energy sources and a shift, hopefully, to plug-in hybrid electric vehicles (PHEVs) and all electric vehicles (EVs) will require a much more dynamic electric infrastructure. Beginning with the U.S. DoE "Grid 2030 Vision" report, energy storage emerged as a top concern for the future. In 2007 the DOE convened an Electricity Advisory Committee (EAC) to make recommendations for an energy road map for the U.S. including energy storage. The EAC produced a report (Department of Energy, 2008) to the U.S. Congress, which provided a road map development of storage technologies and goals for storage deployment in the U.S. grid over a ten-year period. Globally, other nations like Japan and Germany have been working to make larger amounts of energy storage a vital part of their energy plan. Japan has a near-term target of 15% storage in the grid with Germany planning 10% compared to just over 2% in the U.S. (Masiello, 2009).

There are several categories to consider when classifying energy storage units; size and weight, capital costs, and life efficiency. Size and weight of storage devices are important factors for certain applications. The influence of size and weight with respect to output energy density as well as the values of various battery types can be seen in Figure 2.2 (Department of Energy, 2009). However, the electrically rechargeable types, such as zinc-air batteries, have a relatively small cycle life and are still in the development stage. The energy density ranges reflect the differences among manufacturers, product models and the impact of packaging.

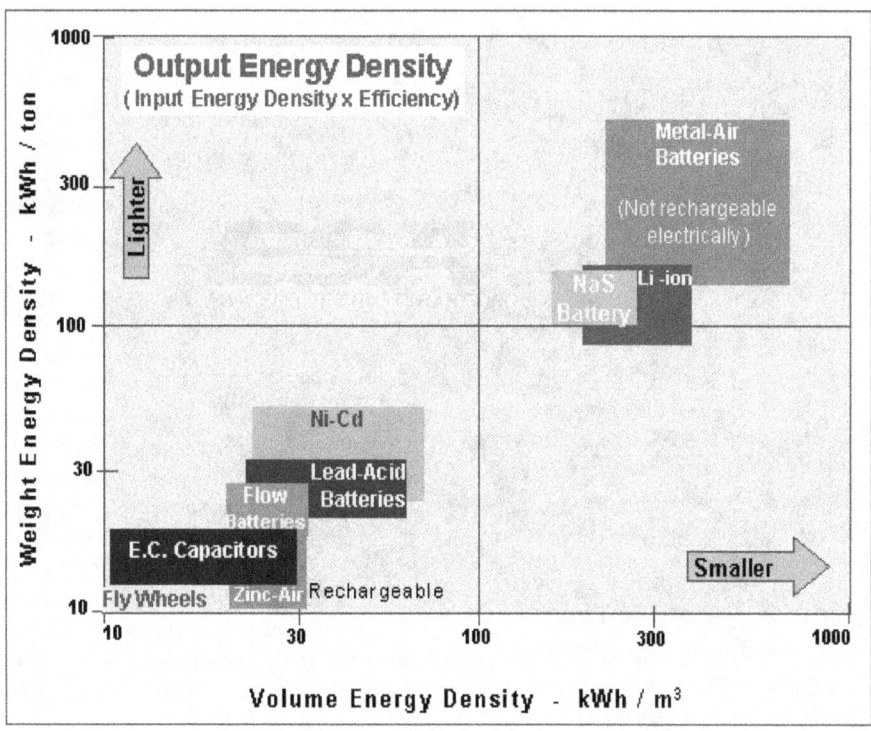

Figure 2.2 *Graph of storage systems based on output energy density*
(Department of Energy, 2009)

While capital cost is an important economic parameter, it should be realized that the total ownership cost is a much more meaningful index for a complete economic analysis. For example, while the capital cost of lead-acid batteries is relatively low (as shown in Figure 2.2) (Department of Energy, 2009), they may not necessarily be the least expensive option for energy management due to their relatively short life for this type of application. The battery costs in Figure 2.3 have been adjusted to exclude the cost of power conversion electronics. The cost per unit energy has also been divided by the storage efficiency to obtain the cost per output (useful) energy. Installation cost also varies with the type and size of the storage. The information in the chart and table here should only be used as a guide not as detailed data.

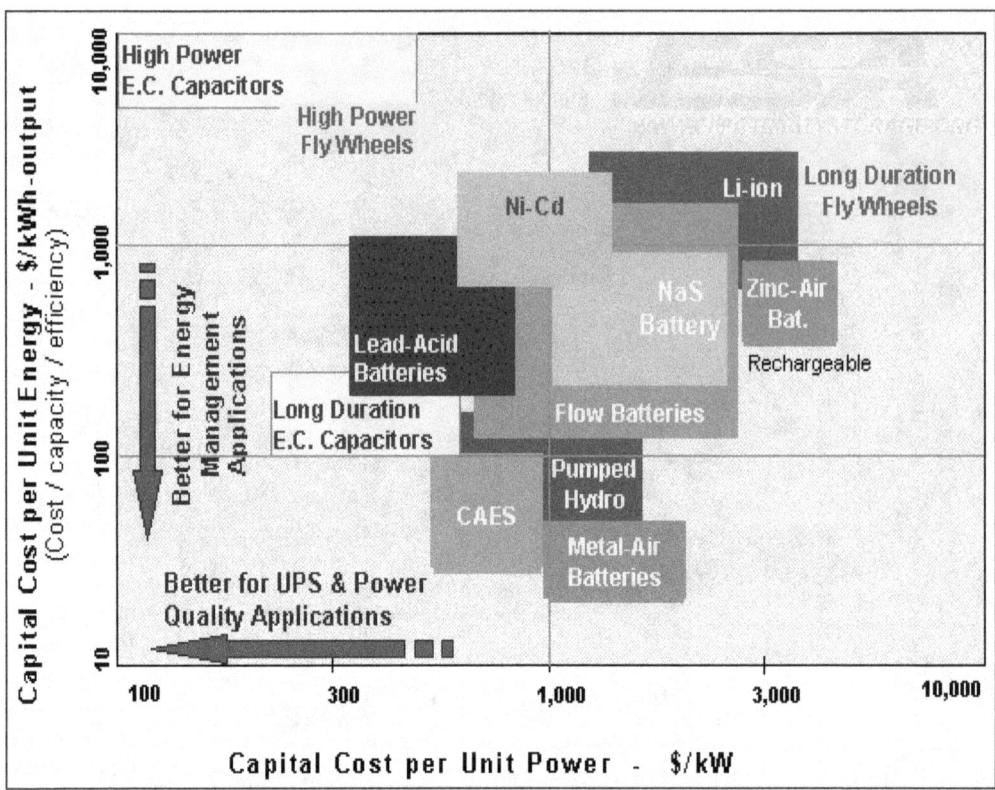

Figure 2.3 Graph of storage systems based on capital cost and power (Department of Energy, 2009)

Efficiency and cycle life are two important parameters to consider along with other parameters before selecting a storage technology. Both of these parameters affect the overall storage cost. Low efficiency increases the effective energy cost as only a fraction of the stored energy could be utilized. Low cycle life also increases the total cost as the storage device needs to be replaced more often. The present values of these expenses need to be considered along with the capital cost and operating expenses to obtain a better picture of the total ownership cost for a storage technology.

Advanced battery technologies have been the main focus in distributed storage systems followed by flywheels, supercapacitors, and compressed-air energy storage (Roberts, 2011). Studies conducted in 2009 to determine the effectiveness of fast-response batteries and flywheels in frequency regulation applications showed these systems could affect frequency control with approximately 40% less energy as compared to fossil fuel plants because of the very fast response time (cycles versus minutes to respond). Based on these findings plants up to 20 MW each are being built in the U.S. and other countries (Rastler, 2008). In the wind turbine industry super-capacitors have been widely adopted for powering the pitch control of turbine blades and offer back-up energy to safely shutdown a wind turbine if loss of power occurs. Super-caps are being applied as well with small solar arrays to insure smooth power flow as clouds pass over.

Lithium-ion batteries, which have achieved significant penetration into the portable/consumer electronics markets and are making the transition into hybrid and electric vehicle applications, have opportunities in grid storage as well. If the industry's growth in the vehicles and consumer electronics markets can yield improvements and manufacturing economies of scale, they will likely find their way into grid storage applications too. Developers are seeking to lower maintenance and operating costs, deliver high efficiency, and ensure that large banks of batteries can be controlled. As an example, in November 2009, AES Energy Storage and A123 Systems announced the commercial operation of a 12 MW frequency regulation and spinning reserve project at a substation in the Atacama Desert, Chile. Continued cost reduction, lifetime and state-of-charge improvements, will be critical for this battery chemistry to expand into grid applications (Doughty et al, 2010).

Most modern high-speed flywheel energy storage systems consist of a massive rotating cylinder (a rim attached to a shaft) that is supported on a stator by magnetically levitated bearings. To maintain efficiency, the flywheel system is operated in a vacuum to reduce drag. The flywheel is connected to a motor/generator that interacts with the utility grid through advanced power electronics. Some of the key advantages of flywheel energy storage are low maintenance, long life (20 years or tens of thousands of deep cycles), and negligible environmental impact. Flywheels can bridge the gap between short-term ride-through power and long-term energy storage with excellent cyclic and load following characteristics. Currently, high-power flywheels are used in many aerospace and UPS applications. Today 2 kW/6 kWh systems are being used in telecommunications applications. For utility-scale storage a 'flywheel farm' approach can be used to store megawatts of electricity for applications needing minutes of discharge duration. Currently several 'flywheel farm' facilities are in the planning or construction stages in order to sell regulation services into open ISO markets (Energy Storage Association, 2011).

Compressed-air energy storage (CAES) uses off peak electricity to compress air into either an underground structure (e.g., a cavern, aquifer, or abandoned mine) or an above ground system of tanks or pipes. The compressed air is then mixed with natural gas, burned, and expanded in a modified gas turbine. In a conventional gas turbine, roughly two thirds of the power produced is consumed in pressurizing the air before combustion. CAES systems produce the same amount of electric power as a conventional gas turbine power plant using less than 40% of the fuel. Recent advancements in the technology include above-ground storage in empty natural gas tanks and 'mini-CAES', a transportable technology that can be installed at or near individual loads (e.g., on urban rooftops). The first commercial CAES was a 290-MW unit built in Hundorf, Germany in 1978. The second commercial CAES was a 110-MW unit built in McIntosh, Alabama in 1991. Several more CAES plants are in various stages of the planning and permitting process (Department of Energy, 2009).

The challenge in development of a more intelligent electricity network (smart grid) is balancing all of the variables associated with dynamic load control powered from an ever increasing variable (renewable energy) sources. This "balancing act" can be made simpler with small amounts of energy stored throughout the grid. A specific example of storage in a smart grid is the concept of placing small amounts of energy storage (1–2 hours) on the feeders of residential areas. American Electric Power applied this idea by developing the community energy storage (CES) concept (American Electric Power, 2010). CES units are placed at the very edge of the grid allowing for ultimate voltage control and service reliability. As more and more sophisticated electronic loads, computers, appliances, etc. are added by customers who demand greater service reliability, new even larger loads will be added randomly in the grid. On top of these changing load patterns more and more solar arrays on roof tops will introduce a growing amount of energy flowing back into the grid when solar generation exceeds the power demand of the specific customers. Today, a neighborhood with a significant number of solar roofs and generates a fair amount of energy that dissipates back into the utility network during solar peak periods. Since the solar peak precedes the customer load peak by two to three hours each work day it is desirable to store that energy or use when the load grows later in the day.

Current renewable energy storage systems have varying degrees of variability and uncertainty, and the output characteristics of the associated technologies vary substantially. However, research in this field is advancing rapidly. It is likely that a solution for small scale renewable energy storage will become a reality within the next few years. For the purpose of this report, it will be assumed that a reliable, compact, cost effective storage system is available to consumers and can be effectively applied to Smart grid systems.

This Page Intentionally Left Blank

Chapter 3
Methodology

3.1 Introduction

This portion of the report will detail the methodology behind the new pricing and buying strategies. Important aspects include the description of the use an interface to collect consumer preferences and its influence on pricing methodology. Furthermore, this portion of the report will explain the use of the mathematical equations and simulation to determine the price of energy consumption, provide greater consumer – producer visibility, and mitigate the inefficiencies in the market.

In the first example, this segment of the technical report will implement the methodology of the entire process using two consumers and two producers. The two consumers will have different preferences which will impact the type of input source that is used as well as the price that can be charged by the producers of these input sources. After describing the basic methodology using this simple example, the next portion of the report will step into the implementation of this methodology within a complex real world example with multiple producers and consumers.

This methodology uses demand side management to match prices with customer preferences which will ensure both consumers and producers benefit. It will also identify inefficiencies in the market that can be capitalized on by new producers. The interactions between the base price, as determined by the market, and customer preferences will influence the price of energy consumption.

The goal of this pricing and buying strategy is to enable emerging micro grid and smart grid technologies in addition to helping shape demand preferences. Additionally, this strategy will indicate what preferences the consumers are willing to pay for, to include: lower prices, carbon level, pollution level, resiliency, locally generated, and quality. A visual depiction of the complete process, for any example, can be seen in Figure 3.1.

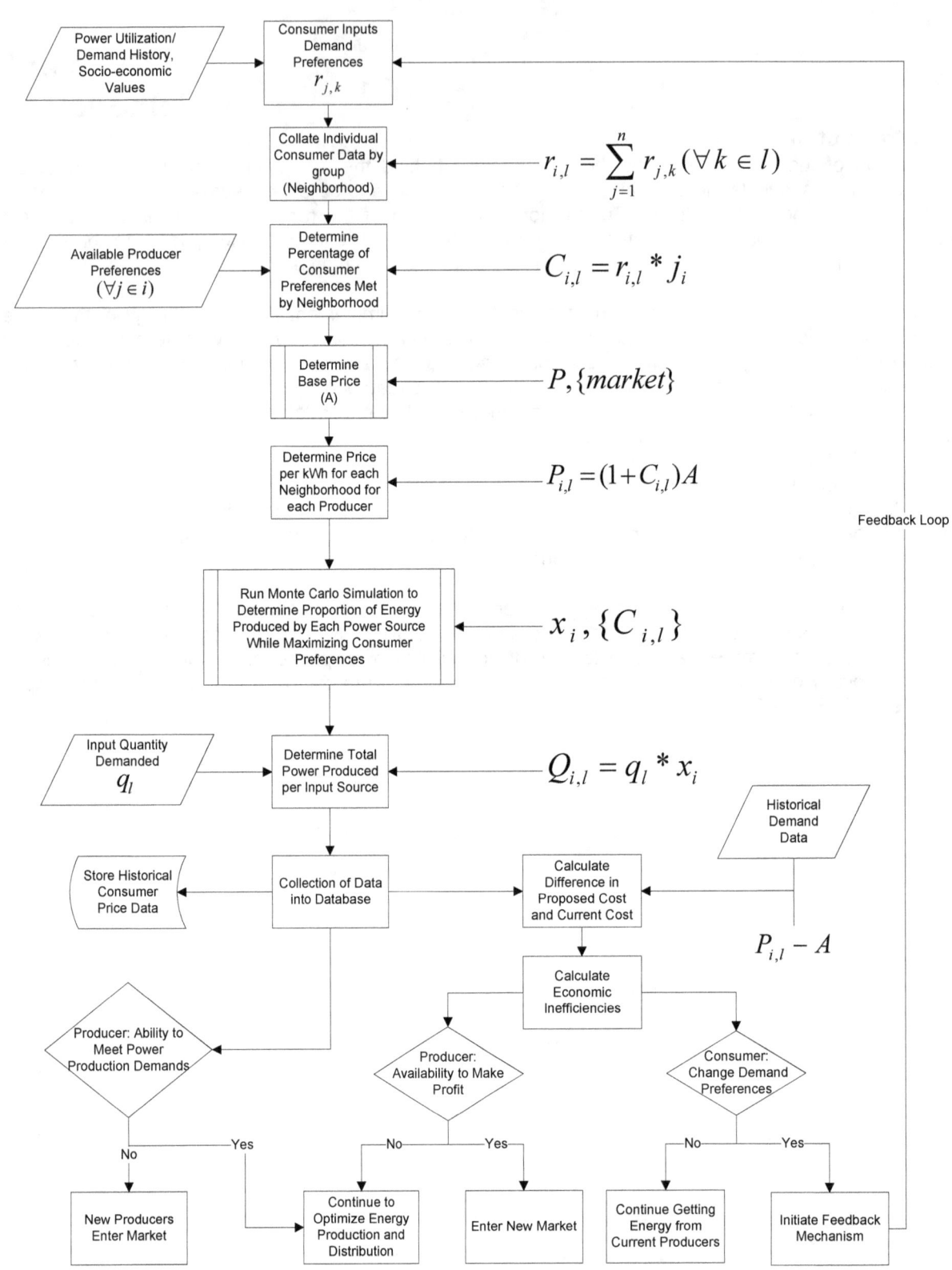

Figure 3.1 *Overall process flow chart*

3.2 Equations

Before stepping into the scenario and simulations it is important to provide an outline of the process and mathematical equations that provide the foundation for the methodology as well as the different parameters. Each of the supporting equations can be seen in Equations 1 - 4.

Equation 1 is used to collate the individual consumer data input from the surveys by group, in this case, neighborhood. This depicts the consumer demand preference profiles for each neighborhood based on the collected data.

$$ \tag{1} $$

The parameters for this equation are as follows:

: Summation of consumer demand preferences for a neighborhood

: Individual consumer demand preference inputs

: For all consumer preferences in a given neighborhood

k : Consumer preferences

: Neighborhood

Equation 2 is used to determine the percentage of consumer preferences met based on the consumer demand preferences for a neighborhood and the available producer capabilities or preferences provided in that neighborhood.

$$ \tag{2} $$

The parameters for this equation are as follows:

: Percentage of consumer preferences met by neighborhood

: Consumer demand preferences by neighborhood

: Available producer capabilities

In the next step, it is necessary to determine the base price, which, for this simulation was assumed to be the market price. Then, Equation 3 is used to determine the price per kWh for each producer in each neighborhood. Willingness to pay for consumer preferences is the driving factor in the difference between this price and the base price.

$$ \tag{3} $$

The parameters for this equation are as follows:

: Consumer preference driven price per kWh

: Percentage of consumer preferences met by neighborhood

: Market price

Next, a Monte Carlo simulation is run in Microsoft Excel to determine the optimal proportion of energy produced by each of the power sources in a given scenario with the goal of maximizing consumer preferences. This value can then be used to determine the total power produced by each power source in Equation 4.

$$ \tag{4} $$

The parameters for this equation are as follows:

Q : Total power produced per input source

: Quantity of power demanded

: Proportion of energy produced per input source (determined through Monte Carlo Simulation)

The following steps include collecting the data and determining the differences in the proposed costs of energy with respect to the market price. This difference can be used to calculate the economic inefficiencies in the market based on this discrepancy in price and the amount of power produced. The methodology, supported by this mathematical background, leads to decisions for both producers and consumers. These decisions serve as feedback mechanisms that reinforce the methodology on a continual basis in order to provide visibility on the optimal mix of power producers to minimize economic inefficiencies with a constant goal of meeting consumer preferences.

3.3 Methodology Walkthrough

The educated and informed customer will submit their individual preferences, in effect, creating a consumer profile. These consumer profiles will include 6 preferences: Carbon Reduction, Pollution Reduction, Improved Reliability, Improved Quality, Renewability, and Local Generation. The consumer can also rank seven input sources: Wind, Coal, Natural Gas, Solar, Hydro, Nuclear, Biomass. The seven power input sources are characterized by the six preferences through a binary table.

Table 3.1 Preference and source table

	Producers						
Preferences	**Wind**	**Coal**	**Nuclear**	**Solar**	**NG**	**Hydro**	**Biomass**
Carbon Reduction	1	0	1	1	0	1	0
Pollution Reduction	1	0	0	1	0	0	0
Improved Reliability	1	1	1	0	1	0	1
Improved Quality	0	1	1	0	1	1	0
Renewable	1	0	0	1	0	1	1
Locally Generated	1	0	0	0	0	0	1

Sources							
Wind	1	0	0	0	0	0	0
Coal	0	1	0	0	0	0	0
Natural Gas	0	0	0	0	1	0	0
Solar	0	0	0	1	0	0	0
Hydro	0	0	0	0	0	1	0
Nuclear	0	0	1	0	0	0	0
Biomass	0	0	0	0	0	0	1

The consumer profiles will capture individual preferences and then will be collated to a neighborhood profile that describes the collective attitudes in a small community. We have named our neighborhood Greyghost to further illustrate the methodology and model. Note that this data is also used in Chapter 4 for our example problem. The consumer preferences for Grehyghost are seen in Table 3.2. In this example, the conditions will dictate that only four power sources can be used: solar, natural gas, biomass, and waste treatment, to illustrate the dynamics of the environment and certain restrictions.

Table 3.2 Consumer preferences for Greyghost neighborhood

	Consumer Preferences by Neighborhood
Carbon	3.67%
Pollution	3.60%
Reliability	3.44%
Quality	1.56%
Renewable	3.63%
Locally Generated	1.38%
Sum	17.28%

This neighborhood profile will be compared with the available producer preferences to determine the percentage of consumer preferences met by the neighborhood. This can be seen in table 3.3.

Table 3.3 Consumer preferences met by various sources in neighborhood profile

Sources	Consumer Preferences Met
Wind	15.72%
Coal	5.00%
NG	0.00%
Solar	10.90%
Hydro	0.00%
Nuclear	8.86%
Biomass	8.45%
Waste Treatment	10.90%

The neighborhood profile will also indicate the amount that consumers would be willing to pay above the market price of energy. Using the market price for energy and the consumer profile, a new price can be determined for each neighborhood for each producer. This can be seen in Table 3.4.

Table 3.4 Source prices for neighborhood profile

Sources	Price
Wind	$ 0.15
Coal	$ 0.14
NG	$ 0.13
Solar	$ 0.14
Hydro	$ 0.13
Nuclear	$ 0.14
Biomass	$ 0.14
Waste Treatment	$ 0.14

This new price coupled with the consumer preferences will be inputted into a Monte Carlo simulation that will determine the proportion of energy produced by each power source while maximizing the consumer preferences. Table 3.5 illustrates the simulation results for neighborhood Greyghost under certain source conditions.

Table 3.5 Optimal energy proportions and consumer preferences met for Greyghost neighborhood

Preferences	Wind	Coal	Nuclear	Solar	NG	Hydro	Biomass	Waste Treatment
Proportion of Energy	0.00%	0.00%	0.00%	3.00%	27.00%	0.00%	40.00%	30.00%
Consumer Preferences Met	0.00%	0%	0.00%	3%	1.08%	0%	34.05%	32.49%

The simulation will be constrained by the realities of the current power grid and technologies available as well as environmental constraints such as wind and solar exposure. Ultimately, the simulation will provide information regarding the total power produced per input source for each neighborhood, a larger population, and that information will then be analyzed to find economic inefficiencies. Table 3.6 illustrates the proposed cot per 30 minute time period for each power source for Neighborhood Greyghost, this can be compared to the original costs to find the total economic inefficiencies for this neighborhood.

Table 3.6 *Proposed source prices for Greyghost neighborhood*

Sources	Cost
Wind	$ -
Coal	$ -
NG	$ 0.42
Solar	$ 3.37
Hydro	$ -
Nuclear	$ 5.41
Biomass	$ 4.15
Waste Treatment	$ -
Sum	$ 13.35
AVG $/kWh (Neighborhood)	$ 0.13907

Table 3.7 illustrates the total economic inefficiencies for neighborhood Greyghost in time periods. The yearly economic inefficiency for this neighborhood is $13,636.39 under the source conditions.

Table 3.7 *Economic inefficiencies for Greyghost neighborhood*

Total Economic Inefficiencies	
30 min	$ 0.83
hour	$1.67
day	$40.06
year	$14,636.39

This information can be used by power suppliers to compare their current situation and whether or not they would want to enter the market to capitalize on the economic inefficiencies that currently impact this neighborhood. With more neighborhoods, this information becomes even more valuable as small microgrids can be established to meet more of the needs and the economic inefficiencies would increase. The process begins with the consumer and ends with a strong statement of economic incentives for increased power investment in different sources for a particular area. This, applied to a macro scale, would create multiple smart and micro grids that capitalize on the market inefficiencies that is not only beneficial for the power providers, but also meets consumer preferences boosting carbon reduction, pollution reduction, increased reliability, quality and bringing in more renewable and locally generated power.

3.4 Summary of Methodology

The methodology behind the new pricing and buying strategy will present an opportunity for energy producers and consumers to optimize electricity usage, meet different consumer preferences and expand the limitations of producers.

As seen through this example, prices vary depending upon consumer demands and the ability of a producer to meet those demands. The demand side management and price based interactions used in this pricing and buying strategy provides many great benefits. By quantifying demand for different input power preferences this approach allows for greater consumer control of energy usage while simultaneously ensuring producers tailor to these specific needs. The increase in consumer- producer visibility serves two purposes: maximizing customer control over preferences and input source in addition to indicating potential demand surpluses and shortages for producers to capitalize on by entering the market or readjusting their production focus. The overarching goal of this approach is to enable emerging technologies such as smart grids and micro grids to greater meet customer preferences as indicated through a willingness to pay prices above the market price. Using price as a driving factor helps to indicate inefficiencies in the market thus increasing new producers in the market. The development of this methodology allows for a simple approach to energy production and consumption while delving into the details of the process to greater meet indicated consumer demands and allows producers to tailor their business to these demands.

The end state outputs of this methodology allow consumers to shape their preferences to attain energy from new input sources using alternative producers. It also forces an informed decision for producers to determine if they have the ability to meet consumers needs which leads to the choice of maintaining the optimal level of operations or entering the market.

This methodology is only the first step toward the future of energy production, distribution, and consumption. The adoption of smart grid technologies as well as the implementation of new infrastructure and technology will further the enhancement and refinement of our energy system and the way it operates. The scope of this project so far is limited, but the methodology and the aggregation of pricing data will pave the way for future work in the energy market.

Chapter 4
Micro Energy Market Study
for the U.S. Military Academy Installation

4.1 Introduction

This chapter details the scenario, simulation, and results for a study of the grid at the United States Military Academy (USMA) at West Point. It will first outline the scenario with the breakdown of the different neighborhood sectors at USMA, then describe the mathematical methodology behind the simulation, and to conclude it will display the results and a detailed analysis of the simulation. This scenario provides an example of the methodology in action for an adequately sized area that captures various preferences and sources.

The USMA case study serves as an example of three different approaches to the energy grid and new pricing and buying strategies. After thorough research and analysis our group determined three different simulation trials: a base scenario or a representation of the current situation at the USMA, a scenario representing the National Renewable Energy Laboratory (NREL) report[5] recommended energy usage for the USMA, and an ideal situation scenario that used all types of energy providers in the grid. Each of these simulations provide data with respect to consumer preferences met as well as the potential amount of money to be earned by each producer over a certain time period. The objective of these scenarios is to compare and contrast the three different potential alternatives for energy distribution and consumption at USMA as an example of a typical Army base or a small community.

Using @Risk[6] Simulation for each scenario, with the data collected about the neighborhoods throughout the USMA campus, the simulations provided data that revealed varying levels of economic inefficiency and opportunity. The key parameters our group focused on include the percentage increase in consumer preferences met for a certain region and the level of economic opportunity available, in dollars, in scenarios with respect to the energy provider.

This methodology and the results provided create a clear picture of the economic inefficiencies present in the current system. The goal of the simulation is to use these results as an impetus for changing pricing and buying strategies to capitalize on these ineffiencies. Focusing on the two key parameters and yielding results that shed light on different investment opportunities with respect to energy consumption and production.

This portion of the report will now describe the USMA case study and its results.

4.2 Scenario

A survey through surveymonkey.com was utilized to capture consumer energy production preferences. Ideally, the app featured in Chapter 3 would be the best way to gather massive consumer information, but for a smaller scale, a survey was used. The survey allowed participants to rank order the six energy types: wind, solar, hydro, natural gas, coal, and nuclear, capturing the percent increases consumers would be willing to invest for energy types. Participants were then asked if they would be willing to pay a premium on varying energy preferences. The energy preferences were:
- Carbon free,
- Pollution free,
- Perfect reliability,

---

[5] Davis, J., Harris, T., Robichaud, R., Tomberlin, G. Hunsberger, R., Scarlata, C., Martin, D., and Huffman, S. *"Targeting NetZero Energy at U.S. Military Academy West Point: Assessment and Recommendations,"* U.S Department of Energy National Renewable Energy Laboratory, June 2012.
[6] @RISK (pronounced "at risk") performs risk analysis using Monte Carlo simulation using spreadsheet models more information can be found at http://www.palisade.com/risk/

- Perfect quality,
- Only renewable, and
- Locally generated.

By filling out this portion of the survey, participants inform producers that they are willing to pay the given % extra premium for a certain energy preference as long as there energy needs are still being satisfied. Figure 4.1 shows the results for a single participant in the survey:

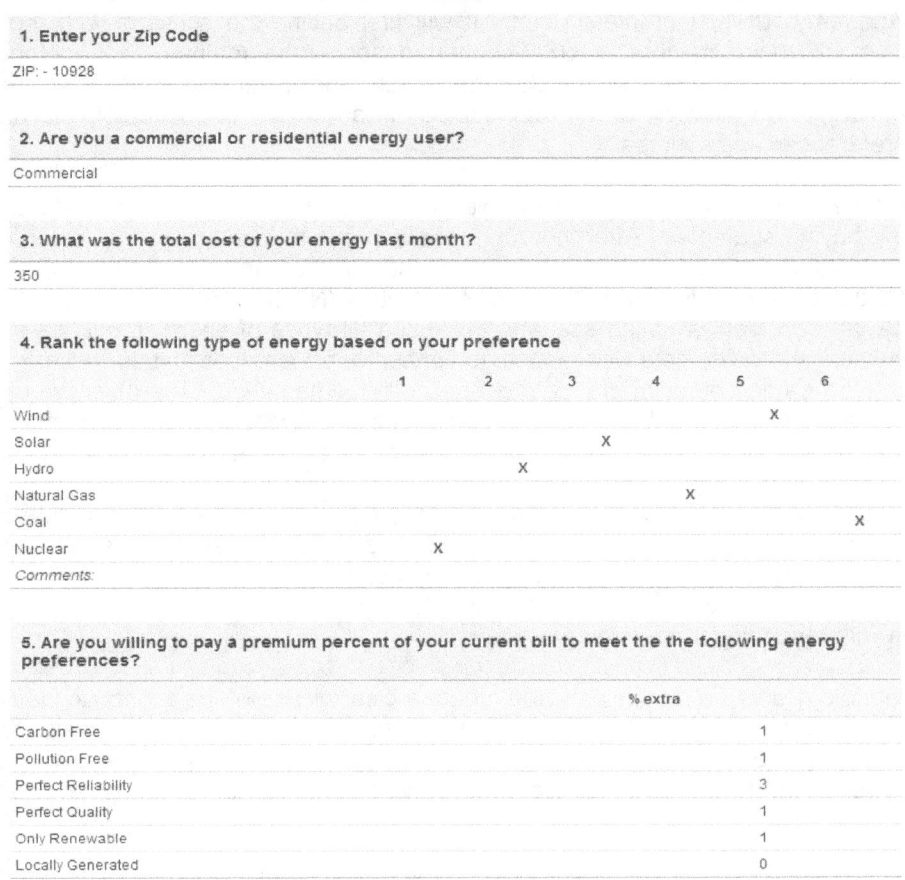

Figure 4.1 *Example survey result*

According to the survey results in Figure 4.1, this participant prefers nuclear energy the most and prefers coal energy the least. They are also willing to pay a premium of 3% of their previous month's price, $360.50, to improve the reliability of their energy.

Considering the scope of our simulation trials it was decided that presenting the survey to faculty members residing in each of the neighborhoods of interest would provide enough data as a representation of the USMA population. The 17 survey participants served as a solid basis for analyzing consumer preferences and pricing in each of the neighborhoods. Based on the 17 survey participants, an aggregate model was created to determine the average preferences for energy type and premiums. The following Excel template was created to capture the aggregate data. For a more robust analysis it would be beneficial to attain more survey results in order to better capture the overall population's preferences and willingness to pay for certain types or qualities of energy.

Table 4.1 *Survey results collection table*

Consumer	Zip	Consumer Type	Cost of Energy	Rank of Preferences						% Extra					
				1	2	3	4	5	6	Carbon Free	Pollution Free	Perfect Reliability	Perfect Quality	Only Renewable	Locally Generated
1	10950	Residential	$400.00	Nuclear	Natural Gas	Hydro	Solar	Coal	Wind	0	0	0.1	0.1	0	0
2	12520	Commercial	$250.00	Natural Gas	Nuclear	Solar	Wind	Hydro	Coal	0		0.05	0	0.03	0
3	12520	Residential	$284.00	Natural Gas	Nuclear	Solar	Coal	Hydro	Wind	0.02	0.05	0.06	0.03	0	0.05
4	10928	Residential	$200.00	Solar	Nuclear	Wind	Hydro	Natural Gas	Coal	0	0.05	0.1	0.05	0.02	0
5	12520	Residential	$150.00	Solar	Wind	Hydro	Nuclear	Natural Gas	Coal	0	0	0	0	0	0
6	10922	Residential	$80.00	Solar	Wind	Hydro	Natural Gas	Natural Gas	Coal	0.2	0.15	0	0	0.1	0
7	10562	Residential	$374.00	Wind	Solar	Hydro	Natural Gas	Nuclear	Coal	0	0	0	0	0.1	0
8	12589	Residential	$336.00	Solar	Wind	Hydro	Natural Gas	Coal	Nuclear	0.1	0.1	0	0	0.05	0.05
9	10093	Commercial	$200.00	Wind	Solar	Hydro	Natural Gas	Coal	Nuclear	0.02	0.02	0.02	0.02	0.02	0.02
10	12589	Residential	$200.00	Nuclear	Hydro	Natural Gas	Wind	Solar	Natural Gas	0.02	0	0.1	0	0	0
11	10036	Commercial	$39.41	Nuclear	Hydro	Natural Gas	Solar	Wind	Coal	0.02	0	0	0	0.05	0
12	10996	Residential	$200.00	Nuclear	Coal	Natural Gas	Wind	Solar	Hydro	0	0	0.02	0	0	0
13	10996	Residential	$150.00	Hydro	Wind	Nuclear	Solar	Natural Gas	Coal	0.1	0.1	0	0	0.05	0
14	10996	Residential	$150.00	Nuclear	Solar	Natural Gas	Coal	Wind	Hydro	0.05	0.05	0.1	0.05	0.1	0.05
15	10996	Residential	$150.00	Solar	Wind	Nuclear	Coal	Natural Gas	Coal	0.02	0.02	0	0	0.05	0.05
16	10996	Residential	$150.00	Hydro	Solar	Wind	Natural Gas	Coal	Nuclear	0	0	0	0	0	0
		AVG	$207.09						AVG	0.0367%	0.0360%	0.0344%	0.0156%	0.0363%	0.0138%

Table 4.1 Survey results collection table (continued)

Sources							Total
Wind	12	25	8	9	4	2	60
Solar	30	20	8	9	4	0	71
Hydro	12	10	20	9	4	2	57
Natural Gas	12	5	16	12	10	1	56
Coal	0	5	0	6	8	8	27
Nuclear	30	15	12	3	2	3	65

Rank Order
Solar
Nuclear
Wind
Hydro
Natural Gas
Coal

Consumer Type	Energy Type
Commercial	Wind
Residential	Solar
	Hydro
	Natural Gas
	Coal
	Nuclear

Table 4.1 displays some critical information. It shows the aggregate premiums for every energy preference and shows the rank ordering of the energy types. This information increases producer-consumer visibility which can help to mitigate some issues that will be discussed later in the report. Based on the NREL report, the percentage increase in energy source preferences were determined. This indicates what the consumer preferences are not only for the aforementioned preferences but also for the energy sources.

Each of the simulation scenarios were constructed in Microsoft Excel. These spreadsheets were then manipulated using @Risk Simulations in Excel. These models required the input of neighborhoods as energy consumers to efficiently capture energy needs at USMA. Figure 4.2 illustrates the 15 different neighborhoods.

Furthermore, the neighborhoods are broken down as shown in Table 4.2.

Table 4.2 USMA energy neighborhoods

Neighborhood	Name
1	Lee
2	Stony
3	Greyghost
4	New Brick
5	Barracks
6	Academics
7	Hospital
8	PX/Commisary
9	Professor's Row
10	Lusk
11	Michie
12	Thayer Road
13	Ike Hall
14	Buckner
15	Rugby Complex

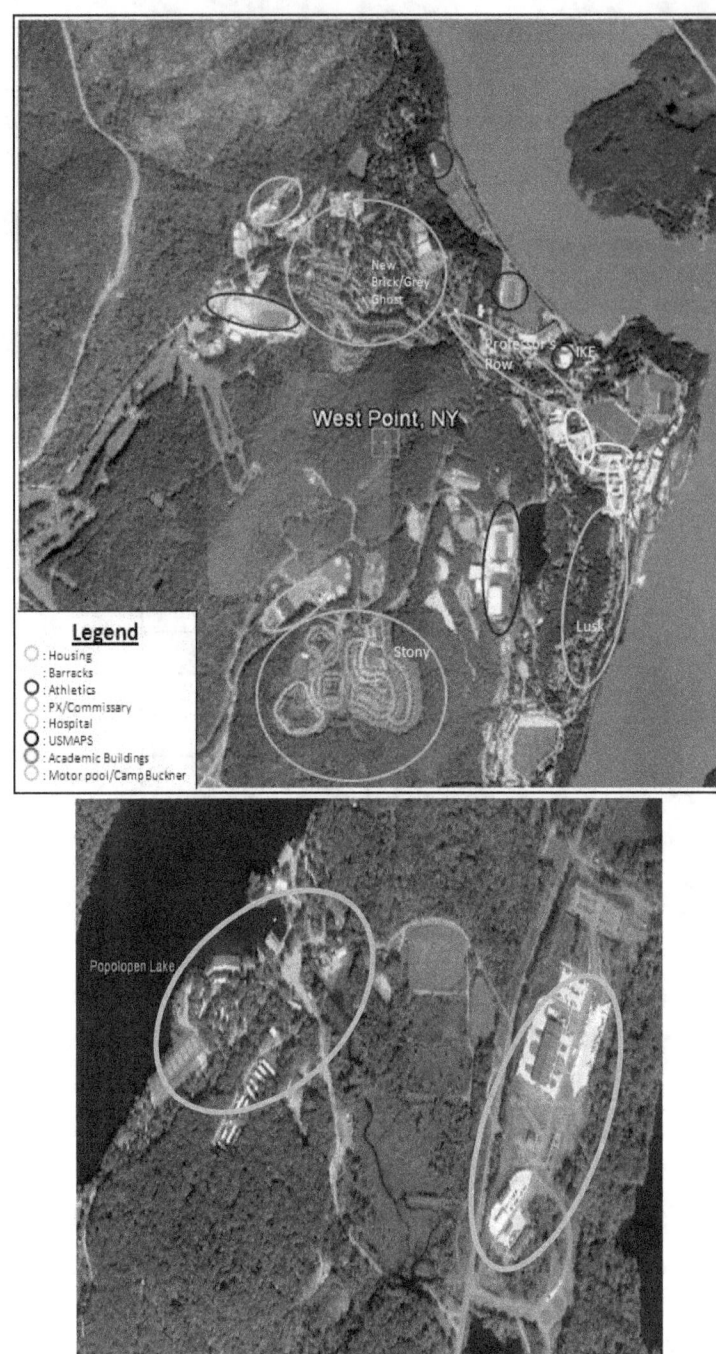

Figure 4.2 Neighborhood diagram

The neighborhood profiles were broken down by the type of facilities or houses, the number of houses, and the size of the houses. Standard values for each type of house were used to create unique neighborhood profiles for each of the 15 neighborhoods at USMA selected for the analysis. This scenario captures the unique neighborhood profiles through surveys, interviews, and the NREL report for USMA. Once the amount of energy usage for each of the neighborhoods was determined, our group used mathematical analysis and simulation techniques to maximize the consumer preferences met by altering the proportion of energy input by each source for a certain neighborhood. The housing energy profiles and energy usage table can be found in Table 4.3

Table 4.3 Housing energy profiles

	Month	Day	Hour	30 min		
1k ft^2	1400.0	46.7	3.9	1.9		
2k ft^2	1502.0	50.1	4.2	2.1		
3k ft^2	1657.0	55.2	4.6	2.3		
Neighborhood	Name	House Type	Energy	Number Houses	Total Energy
N1	Lee	2k ft^2	2.1	60.0	125.2
N2	Stony	2k ft^2	2.1	217.0	452.7
N3	Greyghost	1k ft^2	1.9	64.0	124.4
N4	New Brick	1k ft^2	1.9	76.0	147.8
N9	Professor's Row	3k ft^2	2.3	20.0	46.0
N10	Lusk	1k ft^2	1.9	22.0	42.8
N12	Thayer Road	3k ft^2	2.3	17.0	39.1

The methodology portion of this chapter goes into further detail with regards to the steps taken to accurately model the energy production and pricing strategy at USMA for each of the 15 neighborhoods outlined in this scenario.

4.3 Methodology

In order to determine the proportion of energy to produce and the level of consumer preferences met, the simulation used historical data, consumer preferences derived from a survey as well as some assumptions to create preference distributions and run the simulation. The first step matched up general preferences with producers in a binary system, whereby a preference was matched up with a specific producer with a zero or a one. The survey results provided the consumer group preferences and sources. For example, the Lee Neighborhood desired a 3.63% increase in price for carbon reduction, a 1.56% increase for redundance, etc. The survey results provided consumer preference information only for the neighborhoods while interviews with the Environmental and Energy Officer at USMA lent information in which assumptions could be made on other key areas including the academic buildings and recreational facitilites. Tables 4.4 and 4.5 illustrate the binary system and the neighborhood preferences.

Table 4.4 Binary system

	Wind	Coal	Nuclear	Solar	Natural Gas	Hydro	Biomass
Carbon Reduction	1	0	0	1	0	1	0
Pollution Reduction	1	0	0	1	0	0	0
Improved Reliability	1	1	0	0	0	0	1
Improved Quality	0	1	0	0	0	1	0
Renewable	1	0	0	1	0	1	1
Locally Generated	1	0	0	0	0	0	1

Table 4.5 Neighborhood preferences

	N1	N2	N3	N4	N5	N6	N7	N8	N9	N10	N11	N12	N13	N14	N15
Carbon	3.67%	3.67%	3.67%	3.67%	2.00%	2.00%	2.00%	2.00%	3.67%	3.67%	2.00%	3.67%	2.00%	2.00%	2.00%
Pollution	3.60%	3.60%	3.60%	3.60%	2.00%	2.00%	2.00%	2.00%	3.60%	3.60%	2.00%	3.60%	2.00%	2.00%	2.00%
Reliability	3.44%	3.44%	3.44%	3.44%	3.00%	3.00%	5.00%	3.00%	3.44%	3.44%	3.00%	3.44%	3.00%	3.00%	3.00%
Quality	1.56%	1.56%	1.56%	1.56%	3.00%	3.00%	5.00%	3.00%	1.56%	1.56%	3.00%	1.56%	3.00%	3.00%	3.00%
Renewable	3.63%	3.63%	3.63%	3.63%	0.00%	0.00%	0.00%	0.00%	3.63%	3.63%	0.00%	3.63%	0.00%	0.00%	0.00%
Locally Generated	1.38%	1.38%	1.38%	1.38%	0.00%	0.00%	0.00%	0.00%	1.38%	1.38%	0.00%	1.38%	0.00%	0.00%	0.00%

The consumer groups' preferences are matched up with the binary system to calculate the percentage of consumer preferences met. The result is a percentage increase from market price to determine a new price per kWh a specific neighborhood was willing to pay for each specific energy source. This also allows the findings of the minimum and maximum load constraint from each energy source using optimization software to maximize consumer preferences. Each source's minimum load constraint was 5% to create a more robust solution capturing all the sources. The source's maximum load constraint was capped to represent the feasible maximum load each source could provide to the neighborhoods.

@Risk was chosen for the model due to its optimization and simulation capabilities. Each preference for the neighborhoods was defined as a resample distribution. This would randomly select an input from the survey results and select it as the value for that trial. One thousand trials were run to simulate almost every possible combination of consumer preferences. The total *consumer preference met* as well as the *average price* for each energy source were defined as outputs. After running at @Risk, the model produced graphs for each of the defined outputs. Figure 4.3 contains an output graph distribution for the total consumer preferences met in the Ideal Model:

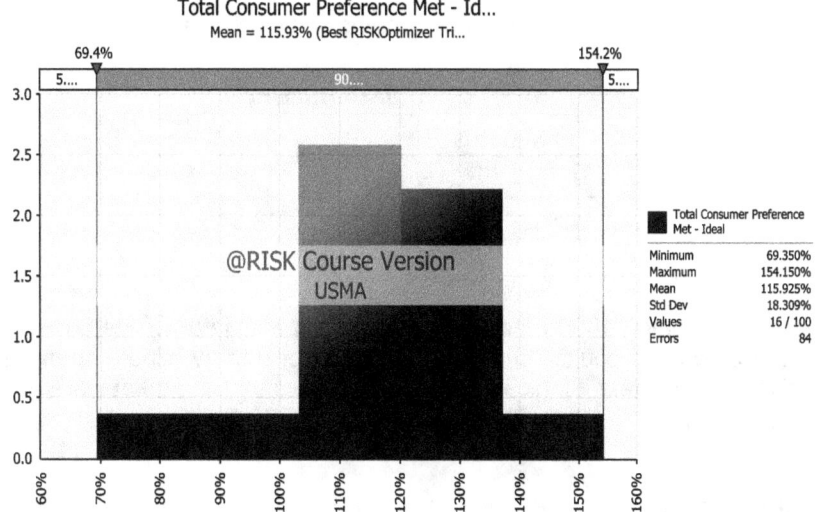

Figure 4.3 Total consumer preference met "Ideal Model"

This output graph serves as an illustrative tool to show the variability of consumer preferences over the thousand trials. A more thorough analysis of the outputs can be found in the analysis portion of this chapter.

Next, using the simulated energy prices per kWh for each energy type compared to the baseline price of $.13 per kWh tables were created to display the discrepancies between the amount of revenue currently generated compared to the potential revenue. In the result section we will present the amount of revenue lost due to consumer preferences not being met in the market for each scenario. Using the "Ideal Model" as an example, the energy producers could make $587,076.60 of untapped revenue if they were able to meet consumer preferences optimally annually at USMA. This serves as a tool to incentivize new investments and new energy producers to enter the energy market at USMA to increase their own revenues while simulatenously increasing the amount of consumer preferences met.

4.4 Results
4.4.1 Baseline
The results from the optimization for the "Baseline Model" are as follows:

Table 4.6 Consumer preference optimization "Baseline Model"

	Wind	Coal	Nuclear	Solar	NG	Hydro	Biomass	*Sum*
Proportion of Energy	0%	0%	70%	0%	30%	0%	0%	*100%*
Total Power Produced (kWh)	0	0	2421.125	0	1037.625	0	0	3458.75
Consumer Preferences Met	0.0%	0.0%	3.5%	0.0%	1.2%	0.0%	0.0%	*4.7%*

The "Baseline Model" is a simulation of the current energy status at USMA in which nuclear and natural gas energy sources are the sole providers. The results of the simulation distributed the production of energy to optimize *consumer preferences met*, to 30% for natural gas and 70% for nuclear. Comparing the new energy price/kWh to the original $0.13, the results shown in Table 4.7 were produced to show the economic inefficiencies that exist due to lack of consumer preferencesmet under this model:

Table 4.7 *Total economic inefficiencies "Baseline Model"*

Total Economic Inefficiencies	
30 Minutes	$ 61.43
Hour	$ 122.86
Day	$ 2,948.54
Year	$ 1,076,218.15

The @Risk simulation modeled 1000 trials. After analyzing the data, we determined that the amount of potential dollars that producers could - but do not - earn over a thirty minute span equals $61.43. When this value is extrapolated our data it indicates that if producers in the market were able to meet consumer preferences they would be able to generate an additional $1,076,218.15 annually.

4.4.2 NREL

The results from the NREL Model optimization are shown in Table 4.8.

Table 4.8 *Consumer preference optimization "NREL Model"*

	Wind	Coal	Nuc-lear	Solar	NG	Hydro	Bio-mass	Waste Treat-ment	Sum
Proportion of Energy	0.00%	0.00%	0.00%	3.00%	27.00%	0.00%	40.00%	30.00%	*100.00%*
Total Power Produced (kWh)	0.00	0.00	0.00	103.76	933.86	0.00	1383.5	1037.62	*3458.7*
Consumer Preferences Met	0.00%	0%	0.00%	3.00%	1.08%	0%	34.05%	32.49%	*70.93%*

Under the NREL model, only energy production suggested by the NREL report were used. The results of the simulation distributed the production of energy to optimize consumer preferences met to 3% for solar energy, 1.08% for natural gas, 34.05% for biomass, and 32.49% for waste treatment energy. Comparing the new energy price/kWh to the original $0.13, Table 4.9 summarizes the economic inefficiencies that exist due to lack of consumer preferences met under this model.

Table 4.9 Total economic inefficiencies "NREL Model"

Total Economic Inefficiencies		
30 Minutes	$	45.49
Hour	$	90.98
Day	$	2,183.63
ear	$	797,023.47

Table 4.9 shows that if producers in the market were able to meet consumer preferences they would be able to generate an additional $772,306.80 annually.

4.4.3 Ideal
The results from the "Ideal Model" optimization are presented in Table 4.10.

Table 4.10 Consumer preference optimizatoin "Ideal Model"

	Wind	Coal	Nuclear	Solar	NG	Hydro	Biomass	Sum
Proportion of Energy	35%	5%	5%	35%	5%	10%	5%	100%
Energy Production (kWh)	1210.56	172.93	172.93	1210.56	172.93	345.87	172.93	3458.75
Consumer Preferences Met	59.85%	5%	0.25%	39%	.2%	11%	4.26%	118.36%

The "Ideal Model" incorporates all of the different potential energy sources to best represent a microgrid. To maximize consumer preferences, the proportion of energy to be produced by coal, nuclear, natural gas and biomass is 5%. Hydro is 10% with wind and solar taking the bulk of the power production with 35%. Comparing the new energy price/kWh to the original $0.13, Table 4.11 displays the economic inefficiencies that exist due to lack of consumer preferences met under this model.

Table 4.11 Total economic inefficiencies "Ideal Model"

Total Economic Inefficiencies		
30 Minutes	$	33.51
Hour	$	67.02
Day	$	1,608.43
Year	$	587,076.60

The above table shows that if producers in the market were able to meet consumer preferences they would be able to generate an additional $587,076.60 annually.

4.5 Analysis
This portion of the report will analyze the three critical outputs of the @Risk simulation models: the energy breakdown, price alterations, and economic inefficiencies. Each of these outputs indicate levels which optimize the efficiency of the energy production cycle in order to maximize the parameter of *consumer preferences met*. Essentially, the percentage of energy produced represents the optimal mix of energy sources and the amount of each energy source that should be produced in a certain scenario.

The price alterations indicate the alterations in prices that can be charged by a producer for a consumer or a group of consumers given their willingness to pay a higher price to meet their preferences. The third output is the most significant, as the economic inefficiencies present in the system indicate the number of dollars lost that could be earned through a better pricing strategy that captures consumer preferences and willingness to pay.

The *consumer preferences met* were determined based on a consumer's willingness to pay percentages of the price higher than the average price for a certain characteristic from their energy source. This willingness to pay serves as a metric to indicate whether or not an energy source is providing what consumers want from their energy. Maximizing this value means that producers can charge more for their product while meeting the needs of its consumers. This metric can be used as a measuring tool to determine the relative level of efficiency within the market and of the different scenarios. A balance should be kept between meeting these preferences and minimizing economic inefficiency. The values for consumer preferences met indicate the percentage above the average price consumers are willing to pay for preferences that the input source can provide. For example, wind power meets the consumer preference of carbon free. The total value of consumer preferences met is simply the sum of consumer preferences met by the various power sources in the model and shows the overall level of compatibility between the mix of energy sources and their ability to meet consumer needs.

Based on the consumer preferences, the optimal mix of energy sources provided for each of the scenarios can be seen in Table 4.12.

Table 4.12 Energy production breakdown by scenario

Proportion of Energy	Wind	Coal	Nuc-lear	Solar	NG	Hydro	Biomass	Waste Treatment	Sum
Baseline	N/A	N/A	70.00%	N/A	30.00%	N/A	N/A	N/A	100.00%
NREL	N/A	N/A	N/A	3.00%	1.08%	N/A	34.05.00%	32.49%	100.00%
Ideal	35.00%	5.00%	5.00%	35.00%	5.00%	10.00%	5.00%	N/A	100.00%

In the previous table, N/A, indicates that the energy sources are not present in the system. For example, in the baseline model, the only energy sources present are nuclear and natural gas. Given USMA's current system the percentage of energy provided by natural gas should be 30% of the required power and the remaining 70% should be provided by nuclear in order to optimize the consumer preferences met. The NREL report recommended that USMA uses solar, natural gas, hydro, biomass, and waste treatment for its energy. In the Ideal model all of the energy sources were available to produce the necessary power. These values indicate the optimal mix and level of production for energy sources. They can be used to indicate to producers how much energy they are required to and should be producing based on consumer preferences. Given these values, an installation or a community could shift its focus to ensure its energy consumption and production more closely mirrors the percentages output from the model in order to best serve the needs of its consumers.

The economic efficiencies were determined by finding the difference between the current price charged for each energy source, $0.13 per kWh, and the calculated cost for each neighborhood and each energy source. The calculated energy cost for each neighboohold was found by summing the the total cost of energy per neighborhood for the 30 minute period and dividing the result by the total power required for that neighborhood. These costs were unique to each neighborhood as they were driven by varying consumer preferences and power requirements. The discrepencies in energy prices allow us to calculate the economic inefficiencies meaning the profit lost per producer every thirty minutes. Table 4.13 illustrates the calculation for the average enegy price per 30 minutes per neighborhood for the "Baseline Model":

Table 4.13 Average energy price calculation per 30 minutes

	Wind	Coal	Nuclear	Solar	NG	Hydro	Biomass	Sum	AVG $/kWh (Neighbor-hood)
N1	$ 4.76	$ 0.62	$ 0.59	$ 4.56	$ 0.59	$ 1.28	$ 0.64	$ 13.02	$ 0.14414
N2	$ 17.22	$ 2.23	$ 2.13	$ 16.50	$ 2.13	$ 4.63	$ 2.30	$ 47.13	$ 0.14414
N3	$ 5.05	$ 0.66	$ 0.62	$ 4.84	$ 0.62	$ 1.36	$ 0.68	$ 13.84	$ 0.14414
N4	$ 5.95	$ 0.77	$ 0.73	$ 5.70	$ 0.73	$ 1.60	$ 0.80	$ 16.29	$ 0.14414
N5	$ 23.44	$ 3.32	$ 3.13	$ 22.79	$ 3.13	$ 6.57	$ 3.22	$ 65.61	$ 0.13624
N6	$ 22.01	$ 3.11	$ 2.94	$ 21.39	$ 2.94	$ 6.17	$ 3.03	$ 61.59	$ 0.13624
N7	$ 20.43	$ 2.95	$ 2.68	$ 19.50	$ 2.68	$ 5.73	$ 2.81	$ 56.77	$ 0.13780
N8	$ 26.29	$ 3.72	$ 3.51	$ 25.55	$ 3.51	$ 7.37	$ 3.62	$ 73.57	$ 0.13624
N9	$ 1.77	$ 0.23	$ 0.22	$ 1.70	$ 0.22	$ 0.48	$ 0.24	$ 4.84	$ 0.14414
N10	$ 1.72	$ 0.22	$ 0.21	$ 1.65	$ 0.21	$ 0.46	$ 0.23	$ 4.71	$ 0.14414
N11	$ 26.29	$ 3.72	$ 3.51	$ 25.55	$ 3.51	$ 7.37	$ 3.62	$ 73.57	$ 0.13624
N12	$ 1.50	$ 0.19	$ 0.19	$ 1.44	$ 0.19	$ 0.40	$ 0.20	$ 4.11	$ 0.14414
N13	$ 2.82	$ 0.40	$ 0.38	$ 2.74	$ 0.38	$ 0.79	$ 0.39	$ 7.90	$ 0.13624
N14	$ 2.82	$ 0.40	$ 0.38	$ 2.74	$ 0.38	$ 0.79	$ 0.39	$ 7.90	$ 0.13624
N15	$ 9.54	$ 1.35	$ 1.27	$ 9.27	$ 1.27	$ 2.68	$ 1.31	$ 26.70	$ 0.13624
Sum	$ 171.62	$ 23.89	$ 22.48	$ 165.93	$ 22.48	$ 47.68	$ 23.47	$ 477.56	
AVG $/kWh (Producer)	0.1418	0.138	0.1300	0.1371	0.130	0.1379	0.135		

Table 4.13 illustrates the average energy price with respect to neighborhood as well as energy type. These calculated values can then be compared against the current market energy price per kWh of $.13. There are a few steps in calculating the economic inefficiencies in the market. The first step is to find differences between the cost of energy using the new price of energy per kilowatt hour and the existing price of energy for each neighborhood per energy source. The summation of all of these differences gives the economic inefficiency for a thirty minute time interval. Expanding this out the a year time frame shows the annual profit lost per producer simply by not meeting consumer preferences. This is shown in the following table for the "Baseline Model".

Table 4.14 Economic efficiency and extrapolation for "Baseline Model"

Total Economic Inefficiencies	
1/2 Hour	$61.43
1 Hour	$122.86
1 Day	$2,948.54
1 Year	$1,076,218.15

The value of the economic inefficiencies or lost profits should provide greater visibility into the need for change within the system and emerging investment opportunities. It also highlights which producers have the most to gain by entering the market and meeting consumer preferences. The economic ineffiencies for the one year time frame for each model can be seen in Table 4.15. These values, paired with the consumer-producer visibility with regards to characteristics preferences can help to drive changes to the ways in which energy is provided and priced.

Table 4.15 *Annual economic inefficiencies*

	Annual Economic Inefficiencies
Baseline	$1,076,218.15
NREL	$797,023.47
Ideal	$587,076.60

4.6 Future Work

For future consideration, our study was limited by a sample distribution of consumer needs at the USMA. A larger distribution of consumer needs would have created more robust simulation factors. Whether through an application for mobile devices or physical means of data gathering, obtaining a larger distribution of consumer needs can create different scenarios to run within the simulation and therefore provide better results in terms of specific energy sources and the potential profit margins that could be available. Furthermore, better stakeholder analysis of the Army's interests and finding the amount that they are willing to invest in the infrastructure itself and pay for energy would have allowed for the development of innovative solutions for energy providers to agree to enter the USMA network or other Army networks.

The implementation of a smart grid with varied power sources would mean the power source providers agreeing to the arrangement. Outreaching to the power providers can provide valuable data on what their magic price point is to enter a new smart grid network and create more tangible options for solution implementation. Furthermore, the power providers can bring expert analysis on external factors such as climate and weather conditions, which are extremely relevant for some renewable power sources. The feasibility of any solution also should be studied, as each solution relies on varying power inputs from differing sources and the overall solution may not be possible under certain environment and policy restrictions and constraints.

Further investigation should identify and harness specific pricing parameters within the simulation to create more realistic and robust pricing simulations to provide to current and potential future producers. Furthermore, a system dynamic simulation modeling with multiple decision making agents could provide an easily reflexive model that can shed insight to second and third-order effects as well as display more detailed results that widens the domain of the scenario.

This study can be used to further research in the implementation of Smart Grids in small and regional areas. Individual consumer preference collection and distribution networks are one of the keys for optimizing networks and minimizing waste while maximizing savings. Though our work does not go deeply into distribution networks and demand response, individual consumer information collection and response can lead to future work and research into those fields. The development of a distributed management system coupled with strategic power generation through load management can also have a critical role in energy pricing strategies and optimization.

Chapter 5
Summary and Conclusion

This report attempted to solve a pressing issue facing the U.S. Army and the nation as a whole. This nation is currenty in a state of dependency on depleting resources attained largely from foreign sources; this raises concerns over energy security and ecological initiatives such as NetZero. With the interactions between reduced budgets, energy security, and NetZero initiatives in mind, this project sought to determine a new pricing and buying strategy for the U.S. Army energy procurement process. To do so, this project analyzed the effects of emerging energy sources and the use of micro grids paired with systemic modifications generating greater producer-consumer visibility to help generate a more efficient means of pricing, buying, and providing energy. The methodology used attempted to determine possible avenues for change in order to best solve this critical problem via simulations.

The process of obtaining the economic inefficiencies within the system starts at first, getting the consumer preferences. Ideally, a consumer-driven application would capture the consumer preferences per household to create a consumer energy profile. This profile would include energy characteristics such as quality, redundancy, and carbon output and energy source preferences such as wind, solar, or coal. This consumer profile would be the starting point to calculate the economic inefficiencies present in the system and furthermore, the groundwork to examine individualized distribution network. The individual consumer preferences were grouped into neighborhood and the consumer preferences for both preferences and sources were derived for each neighborhood. In order to best meet the consumer preferences while adjusting prices, a Monte Carlo simulation was run to maximize consumer preferences by neighborhood while balancing price to determine the amount of power each source should provide for each neighborhood. The new prices per neighborhood were compared to the current prices to determine the total economic inefficiencies, or how much is lost because consumer preferences were not met.

Ultimately this report would provide the basis for meeting individual consumer preferences while fluctuating costs to determine the amount of lost revenue. This lost revenue could be used to create a more robust power grid like a smart grid. Furthermore, energy security and other factors could also be improved using this lost revenue. The tradition hub and spoke model of energy distribution is antiquated and its weaknesses were invariably exposed by natural disasters such as Hurricane Sandy. The implications of such a weak power distribution network are vast. In order to strengthen the network and protect it from future natural disasters and attack, a fundamental change in how power is produced and priced must take place.

The results of the @Risk simulation suggest proportions of certain types of energy to be produced in order to optimize consumer preferences and capitalize on economic inefficiencies. In the baseline scenario, 30% of USMA's energy should come from nuclear and 70% from natural gas. For the NREL scenario, 3% should be Solar Energy, 27% Natural Gas, 40% Biomass, and 30% Waste Treatment Energy. Lastly, for the Ideal scenario, coal, nuclear, natural gas and biomass should each produce 5%, while hydro would produce 10% and wind and solar would each produce 35%.

Based on our simulations, our group was able to model three separate scenarios and determine possible energy production systems that best meet consumer preferences. Meeting consumer preferences was the driving force behind the proportions of energy provided and the price charged for each source. The difference between the current price charged for an energy source and the calculated cost that an energy provider could charge indicates investment opportunities and possibilities to make greater profit while meeting the needs of the consumer. The values found can be used to shed light on the need for new pricing and buying strategies to be applied to various energy profiles in order to increase economic efficiency and quell budgetary concerns.

This report should serve as a starting point for future research with regards to pricing and buying strategies with regards to micro grid technology and emerging energy sources. The scope of our research, while limited, provided a basis to test and implement a methodology that prioritized diversifying energy sources in order to best meet consumer preferences for which consumer will pay more. More detailed and in depth research into this topic is imperative as the budgetary concerns, resource constraints, and the importance of energy security will only increase in severity over time. Some of the topics that should be researched further include: micro grid and alternative energy investment opportunities, consumer preference and demand response pricing, and increasing visibility between consumers and producers in the energy sector.

Chapter 6
References

Banks, Lois. *"Why are New York City's Electricity Rates so High?,"* Federal Reserve Bureau New York, accessed on 24 August 2012 at http://www.newyorkfed.org/research/quarterly_review/1984v9/v9n1article9.pdf 1984

Braun, Martin and Strauss, Philip. *"A Review on Aggregation Approaches of Controllable Distributed Energy Units in Electrical Power Systems",* International Journal of Distributed Energy Sources, accessed on 30 August 2012 at http://www.iset.uni-kassel.de/abt/FB-A/publication/2008/2008_Der_Journal_Strauss_Braun.pdf 17 June 2008

Bureau of Labor Statistics. *"Average Energy Prices in New York- Northern New Jersey,"* accessed on 20 August 2012 at http://www.bls.gov/ro2/avgengny.pdf 25 July 2012

Chang, Chun Ping and Berdiev, Aziz. *"The political economy of energy regulation in OECD countries",* Energy Economics, Volume 33, Issue 5, September 2011, Pages 816-825 http://www.sciencedirect.com/science/article/pii/S014098831100123X

Davis, J., Harris, T., Robichaud, R., Tomberlin, G. Hunsberger, R., Scarlata, C., Martin, D., and Huffman, S. *"Targeting NetZero Energy at U.S. Military Academy West Point: Assessment and Recommendations,"* U.S Department of Energy National Renewable Energy Laboratory, June 2012

Department of Energy. *"2010 Smart Grid System Report Report to Congress,"* accessed on 19 February 2013 at http://energy.gov/sites/prod/files/2010%20Smart%20Grid%20System%20Report.pdf 8 November 2012

Department of Energy. *"Grid 2030: A National Vision for Elecricity's Second 100 Years,"* accessed at http://energy.gov/oe/downloads/grid-2030-national-vision-electricity-s-second-100-years on 19 February 2013

Department of Energy. *"Storage Technology Comparison, Energy Storage Association,"* accessed on 19 February 2013 at http://www.electricitystorage.org/technology/storage_technologies/technology_comparison, April 2009.

Department of Energy. *"Bottling Electricity: Storage as a Strategic Tool for Managing Variability and Capacity Concerns in the Modern Grid,"* DOE Electricity Advisory Committee, accessed on 24 August 2012 at www.doe.energy.gov/eac 2008

Doughty, Daniel, Butler, Paul, Akhil, Abbas, Clark, Nancy, and Boyes, John. *"Batteries for Large- Scale Stationary Electrical Energy Storage,"* accessed on 25 August 2012 at http://www.electrochem.org/dl/interface/fal/fal10/fal10_p049-053.pdf, 2010

Eto, J., Budhraja, V., Martinez, C., Dyer, J., and Kondragunta, M. *"Research, development, and demonstration needs for large-scale, reliability-enhancing, integration of distributed energy resources,"* System Sciences, Proceedings of the 33rd Annual Hawaii International Conference 4-7 January 2000

Fahey, Jonathan. *"Electric Deregulation Finally Takes Off. Forbes USA."* Forbes, 12 Apr 2012. Accessed at http://www.forbes.com/forbes/2010/0412/outfront-electricity-deregulation-constellation-power-moves.html on 24 August 2012

Fan, Jiyuan. *"The Evolution of Distribution. Power and Energy Magazine IEEE,"* 24 August 2012. Vol. 7, No. 2, pp. 63-68, March-April 2009

Faruqui, Ahmad and Sergici, Sanem. *"Household Response to Dynamic Pricing of Electricity – A Survey of the Empirical Evidence."* Brattle Group, pp. 1- 59, February 2010

Faruqui, Ahmad and Sergici, Sanem. *"Dynamic Pricing: What Have we Learned?,"* Brattle Group, 19 May 2011

Fuller, J. C., Prakash Kumar, N., and Bonebrake, C.A. *"Evaluation of Representative Smart Grid Investment Grant Project Technologies: Demand Response"* U.S. Department of Energy, February 2012

Gerlagh, Reyer. *"A climate change policy induced shift from innovations in carbon-energy production to carbon-energy savings",* Energy Economics, Volume 30, Issue 2, March 2008, Pages 425 – 448 http://www.sciencedirect.com/science/article/pii/S0140988306000739

Guldmann, Jean-Michael. *"A Marginal-Cost Pricing Model for Gas Distribution Utilities,"* Operations Research, Vol. 34, No. 6), pp. 851-863, INFORMS, http://www.jstor.org/stable/170766, November - December, 1986

Hernandez-Aramburo, C.A., Green, T.C., Mugniot, N. *"Fuel consumption minimization of a microgrid,"* IEEE Transactions on Industry Applications, Vol. 41, No.3, pp. 673- 681, doi: 10.1109/TIA.2005.847277, May-June 2005

Ipakchi, A. *"Grid of the Future. Power and Energy Magazine,"* IEEE, Vol. 7, No. 2, pp. 52-62, March-April 2009

Kiesling, Lynne. *"Smart Grid: Dynamic Pricing is Smart Grid's Secret Sauce,"* Smart Grid News, accessed on 24 August 2012 at http://www.smartgridnews.com/artman/publish/article_441.html, 13 May 2008

Larsen, E.R. and Bunn, D.W. *"Deregulation in Electricity: Understanding Strategic and Regulatory Risk,"* The Journal of the Operational Research Society, Vol. 50, No. 4, 1999

Marnay, C. and Venkataramanan, G. *"Microgrids in the evolving electricity generation and delivery infrastructure,"* Power Engineering Society General Meeting, IEEE , Vol. ?, No. ?, pp.5 pp., ?-?, 2006

Masiello, Ralph. *"Bottling Energy Issue,"* IEEE Power Energy Magazine, Vol. 7, No. 4, accessed on 19 Febuary 2013 at http://ieeexplore.ieee.org/stamp/stamp.jsp?tp=&arnumber=5159602 July/August 2009

Nogales, F.J. and Conejo, A.J. *"Electricity Price Forecasting through Transfer Function Models,"* The Journal of the Operational Research Society, Vol. 57, No. 4 pp. 350-356, Palgrave Macmillan Journals, http://www.jstor.org/stable/4102386, April 2006

Obama, Barack. *"Energy, Climate Change, and our Environment",* accessed on 27 February 2013 at www.whitehouse.gov/energy 26 February 2013

Paul, Anthony, Palmer, Karen, and Woerman, Matt. *"Modeling a clean energy standard for electricity: Policy design implications for emissions, supply, prices, and regions",* Energy Economics, Volume 36, March 2013, Pages 108 – 124. http://www.sciencedirect.com/science/article/pii/S0140988312003234

Rahimi, Farrokh and Ipakchi, Ali. *"Demand Response as a Market Resource Under the Smart Grid Paradigm,"* IEEE Transactions on Smart Grid, Vol. 1, No., pp. 82-88, 1 June 2010

Rastler, Dan. *"New Demand for Energy Storage,"* Electric Perspectives: Beacon Power, accessed on 25 August 2012 on http://disgen.epri.com/downloads/2008-09-01-EPEnergyStorage.pdf, September-October 2008.

Ramchurn, Sarvapali, Vytelingum, Perukrishnen, Rogers, Alex and Jennings, Nicholas. *"Putting the 'Smarts' into the Smart Grid: A Grand Challenge for Artificial Intelligence,"* Communications of the ACM, Vol. 55, No. 4, pp. 86-97, April 2012

Roberts, Brad. *"Energy Storage Activities in the United States Electricity Grid,"* Electricity Advisory Committee, accessed on August 25 2012 on http://www.doe.gov/sites/prod/files/oeprod/DocumentsandMedia/FINAL_DOE_Report-Storage_Activities_5-1-11.pdf, May 2011.

Sathyanarayana, Bharadwaj. *"Sensitivity-based Pricing and Multiobjective Control for Energy Management in Power Distribution Systems,"* Arizona State University, pp. 1-126, June 2012

Spector, Yishaj, Tishler, Asher and Je, Yinyu. *"Minimal Adjustment Costs and the Optimal Choice of Inputs under Time-of-Use Electricity Rates,"* Management Science, Vol. 41, No. 10, pp. 1679-1692, INFORMS, http://www.jstor.org/stable/2632746, October 1995